I0488146

NUREG-1788

CFD Analysis of Full-Scale Steam Generator Inlet Plenum Mixing During a PWR Severe Accident

Manuscript Completed: May 2004
Date Published: May 2004

Prepared by
C.F. Boyd, D.M. Helton, K. Hardesty

Division of Systems Analysis and Regulatory Effevtiveness
Office of Nuclear Regulatory Research
U.S. Nuclear Regulatory Commission
Washington, DC 20555-0001

ABSTRACT

Computational fluid dynamics (CFD) is used to predict steam generator inlet plenum mixing during a particular phase of a severe accident in a pressurized-water reactor. Boundary conditions are obtained from SCDAP/RELAP5 predictions of a TMLB' station blackout. Full-scale CFD predictions are completed for the scaled-up geometry of a $1/7^{th}$ scale test facility to isolate the scaleup effect. These predictions are repeated with a Westinghouse model 44 steam generator design. The effect of tube leakage on the mixing is also considered. Finally, predictions are completed for a steam generator from a Combustion Engineering (CE) nuclear power plant. Scaleup predictions indicate that data at $1/7^{th}$ scale are indicative of the full-scale behavior for similar geometries. Predictions for a model 44 steam generator design indicate slightly less mixing and increased plume oscillations and indicate that the geometry is an important parameter. Tube leakage does not show a significant impact on the mixing for leakage rates below 1.4 kg/s at these severe accident conditions. A CE steam generator design results in significantly less inlet plenum mixing. The highest tube entrance temperatures approach the hot leg temperatures in this case. Heat transfer rates to the secondary side are determined to be a dominant governing parameter.

CONTENTS

CONTENTS (continued)

Figures

CONTENTS (continued)

CONTENTS (continued)

Tables

EXECUTIVE SUMMARY

This report describes a significant step in the analysis of steam generator inlet plenum mixing using computational fluid dynamics (CFD).

Inlet plenum mixing is one of the parameters governing the degree of thermal challenge to steam generator tubes during postulated severe accidents in pressurized-water reactors (PWRs). The NRC has implemented a steam generator action plan to confirm the robustness of risk-informed licensing decisions and to reduce modeling uncertainties and improve the technical basis for future licensing requests. One objective of this plan is to investigate the time-dependent thermal-hydraulic conditions in the hot leg and steam generator. This research will ultimately lead to a reduction in the uncertainties in modeling these conditions. One aspect of this research involves using state-of-the-art CFD techniques to predict mixing in the steam generator inlet plenum during a particular phase of a TMLB' station blackout event. In an earlier report, the validity of the method was demonstrated through a comparison of predictions with existing experimental data at $1/7^{th}$ scale. The lessons learned from that previous work are utilized to make predictions of full-scale steam generator inlet plenum mixing under a variety of conditions.

The inlet plenum mixing of interest occurs during a station blackout transient where the production of steam and natural circulation flows are the principal means of core cooling. The water level drops below the hot legs and eventually to the bottom of the fuel. With the loop seals plugged with water, the steam flows entering the hot leg and steam generator tubes must ultimately return to the vessel through the same hot leg. A counter current natural circulation flow pattern is established. The hot flow from the vessel travels along the top of the hot leg, through the inlet plenum of the steam generator, and up into a portion of the tubes. Once exiting the tubes into the outlet plenum of the steam generator, the flow returns to the inlet plenum through the remaining tubes and flows back towards the vessel on the bottom of the hot leg. The mixing and entrainment in the inlet plenum reduce the temperature of the hot flow passing through the hot leg from the vessel before it enters the tubes. Significant mixing in the inlet plenum reduces the temperature reaching the tubes, making it less likely they will fail.

The degree of mixing in the steam generator inlet plenum is not computed directly in system analysis codes such as SCDAP/RELAP5 or MELCOR. Inlet plenum mixing is adjusted in these codes to match accepted experimental results or predictions of mixing from other sources. Our understanding of this behavior comes primarily from a set of experiments in a $1/7^{th}$ scale Westinghouse test facility. Questions have been raised concerning the validity of this data over the wide range of applications of interest to the NRC. One question relates to the scale of the experiments and whether these data are applicable at full-scale conditions. Another issue not covered by the experiments is the effect of steam generator tube leakage on the mixing. In addition to these issues, this analysis looks at the significance of steam generator secondary side heat transfer rates and the effect of geometrical variations in the inlet plenum design.

A scaleup analysis is completed that uses a full-scale steam generator with geometry similar to the $1/7^{h}$ scale facility. These predictions indicate that the $1/7^{th}$ scale facility does a good job of representing the full-scale behavior when similar heat transfer rates are applied. Sensitivity studies indicated that the heat transfer rate has a significant impact on the predicted mixing and entrainment in the inlet plenum.

Further analysis using the same boundary conditions applied to the geometry of a model 44 steam generator provides the impact of this inlet plenum geometry on the inlet plenum mixing. The flow patterns are qualitatively different in the nonsymmetric model 44 design compared to the symmetric test facility. Oscillations of the inlet plenum plume spread the region affected by the hot plume over a larger area of the tube sheet and reduce the net average temperature applied to the hottest tubes. The model 44 design places the hot leg nozzle closer to the tube sheet entrance than the design of the 1/7th scale facility. This results in a slightly reduced mixing fraction for the model 44 design compared to the scaleup analysis. Generally speaking, the mixing parameters for the model 44 design are similar to those measured in the 1/7th scale tests.

Tube leakage from a variety of locations is modeled and a few general conclusions are reached. Tube leakage up to 1.4 kg/s (based loosely on a 100 gpm leak size at standard operating conditions) does not significantly impact the inlet plenum mixing. The hot plume is not predicted to divert into the leak and the leak does not cause a bypass (with no mixing) of the inlet plenum. At a leak rate of 2.8 kg/s, the hot plume enters the inlet plenum with less resistance and intersects the tube sheet farther from the hot leg nozzle. The reduced resistance and slightly reduced mixing predicted for the leakage rate of 2.8 kg/s result from the reduced return flow to the vessel. An increase in the maximum tube entrance temperature is predicted for this leak rate. Predictions for leak rates of 0.014 and 0.14 kg/s indicate no significant impact on the mixing parameters.

A steam generator from a Combustion Engineering (CE) plant is considered to study the impact of this type of inlet plenum geometry. The geometry of the inlet plenum for this design places the hot leg relatively close to the tube sheet entrance. Significantly less mixing and entrainment is predicted for the CE design than for the Westinghouse predictions. The highest temperatures entering the tube sheet are very close to the hot leg temperatures. A portion of the hot leg flow enters the tube sheet with little or no mixing. The impact of this reduced mixing on the tube integrity for a CE plant will have to be reevaluated in light of these predictions.

These state-of-the-art predictions provide a detailed look at the flow and mixing in steam generator inlet plenums during severe accident natural circulation conditions. These predictions build upon and extend the range of applicability for the 1/7th scale experiments and provide new insights into conditions beyond the scope of the experiments. In addition, the results provide a more detailed description of the tube temperatures and conditions needed for a tube failure analysis. The predictions should be used with a full understanding of the assumptions and limitations of the approach as outlined in Section 4 of this report.

FOREWORD

Steam generator tube integrity during severe accidents is a critical safety issue. Tube failure during such accidents results in a containment bypass with an associated radioactive release to the environment. Failures of other reactor coolant system (RCS) components before tube failure lead to a depressurization of the RCS and eliminate the threat to the tubes. Existing predictions indicate that the time for RCS failure at locations such as the hot leg or surge line connection is very close to (a few minutes before) the predicted time of tube failure. It is necessary to identify and reduce uncertainties in these predictions to improve the agency's ability to assess the likelihood of steam generator tube failures during severe accidents.

Thermal-hydraulic analysis of the RCS provides the temperature and pressure conditions that challenge the RCS components and the steam generator tubes. The temperature of the steam entering the tubes is influenced significantly by three-dimensional mixing and entrainment in the steam generator inlet plenum region. System codes such as SCDAP/RELAP5 and MELCOR account for this mixing through the use of coefficients predetermined from a set of experiments performed by Westinghouse. The experimental data are valuable but do not answer all of the questions of interest to the NRC. For instance, the effects of tube leakage and inlet plenum geometry variations on the inlet plenum mixing are not considered in the experiments. The Office of Research (RES) has been using state-of-the-art computational fluid dynamics (CFD) to study a variety of safety issues. CFD provides a tool for predicting the steam generator inlet plenum mixing and entrainment under a wide variety of conditions. An assessment of the technique using the existing 1/7[h] scale data as a benchmark demonstrates that the method is capable of predicting the inlet plenum mixing parameters of interest (Ref. NUREG 1781).

This report describes the completion of a detailed analysis of steam generator inlet plenum mixing at various full-scale conditions using CFD. This work extends the results from the 1/7th scale tests to full-scale conditions of interest to the NRC. The analysis supports the September 7, 2000 NRR user need request related to steam generator severe accident response. This request and subsequent related issues are incorporated into the agency's Steam Generator Action Plan (memorandum from Samuel Collins and Ashok Thadani to William Travers, May 11, 2001). This plan is intended to confirm the robustness of risk-informed licensing decisions and to reduce modeling uncertainties and improve the technical basis for future licensing requests.

The successful completion of this work represents an important milestone in the thermal-hydraulic analysis of a reactor coolant system during a severe accident. The predictions provide significant results for full-scale inlet plenum mixing in a Westinghouse and Combustion Engineering (CE) steam generator including the effect of tube leakage. The predictions indicate that the design features of the hot leg nozzle and inlet plenum region have a significant impact on the inlet plenum mixing. The updated mixing parameters outlined in this report are already being incorporated into revised SCDAP/RELAP5 analyses for a Westinghouse and Combustion Engineering plant. Assumptions and limitations of this analysis are outlined in section 4 of this report and should be considered when applying the results.

Farouk Eltawila, Director
Division of Systems Analysis and Regulatory Effectiveness
Office of Nuclear Regulatory Research

1 INTRODUCTION

The Nuclear Regulatory Commission (NRC) has implemented a steam generator action plan[1] to study steam generator tube integrity. This plan includes evaluating the risk of temperature-induced tube rupture during severe reactor accidents. One aspect of this plan is supported by the use of computational fluid dynamics (CFD) to compute steam generator inlet plenum mixing. The transient sequence of interest in this study, a TMLB', begins with a station blackout and ultimately leads to a loss of secondary side cooling and a loss of primary inventory. As the core is uncovered, heat is transferred from the fuel to the metal mass of the primary coolant system through a process of natural circulation. Superheated steam and hydrogen carries heat to structures, including the upper reactor vessel, the hot leg and inlet plenum, and the steam generator tubes. In the specific scenario considered, the loop seals remain filled with water and full loop circulation is blocked. A countercurrent natural circulation flow pattern is expected (and experimentally observed) during this phase of the accident. Figure 1 illustrates this flow pattern.

The scenario ultimately leads to a failure in the primary coolant loop. The thermal-hydraulic details are needed to help determine whether this failure occurs in the reactor coolant piping (in containment) or in the steam generator tubing (with a leak path outside of containment). The fluid mixing phenomena in the steam generator inlet plenum play a significant role in determining the temperature of the steam that reaches the tubes. A lack of mixing in the inlet plenum allows high-temperature steam to enter the tubes, leading to an earlier prediction of tube failure. Alternatively, more complete mixing of the hot steam entering the steam generator with the cooler steam returning to the hot leg reduces the temperature of the steam entering the tubes. This reduces the chance of tube failure, making it more likely that some other component in the system will fail first. If another component fails (for example, the surge line), the system will depressurize into the containment and the threat to the tubes is eliminated. A tube rupture represents a bypass of containment and a potential radioactive release to the environment. Therefore, it is important to accurately predict when the tubes (and other components) are expected to fail. Inlet plenum mixing is one of the parameters governing this tube integrity issue. Background information on severe-accident-induced tube ruptures can be found in NUREG/CR-6285[2] and NUREG-1570.[3]

The thermal-hydraulic modeling of this accident scenario is typically performed with lumped parameter codes such as SCDAP/RELAP5 or MELCOR. The efficiency gained by the coarse nodilization of this approach makes it feasible to predict the behavior of the entire reactor coolant system over extended periods of time. A limitation of this approach is a reliance on predetermined flow-field and mixing parameters to support the prediction of the countercurrent natural circulation phase of the transient. Steam generator inlet plenum mixing and other flow characteristics are obtained from experimental data. The primary source of this data is from a Westinghouse 1/7th scale test facility.

The available test data provide valuable information on steam generator inlet plenum mixing. However, test data is not available over the full range of potential conditions of interest to the NRC. For example, the effect of significant tube leakage on inlet plenum mixing has not been experimentally investigated. Another significant issue relates to inlet plenum geometry. Some steam generators, the Combustion Engineering designs for example, have a significantly different inlet plenum design than typical Westinghouse designs. These design differences are expected to impact the inlet plenum mixing phenomena.

The NRC is using CFD analysis to make predictions over a wide range of full-scale conditions. CFD predictions provide valuable insights into the three-dimensional fluid dynamics and mixing without the expense of testing at these extreme severe accident conditions. A degree of confidence in the technique has been established at 1/7th scale using some of the available test data.[4] The confidence in the technique and the lessons learned at 1/7th scale are carried forward and applied at full-scale conditions. The completed predictions provide valuable insights into the mixing behavior under a variety of full-scale conditions.

The FLUENT (version 6.0) CFD code is used to predict the inlet plenum mixing and the natural circulation flows. FLUENT is a commercially available, general-purpose CFD code capable of solving a wide variety of fluid flow and heat transfer problems. The code solves the Reynolds-averaged Navier-Stokes equations on a finite volume mesh. The Navier-Stokes equations represent the mass, momentum, and energy conservation equations for a continuous fluid. Reynolds averaging creates the need for turbulence modeling to account for the turbulent diffusion of momentum and energy. The FLUENT code provides several turbulence modeling options. Unstructured meshing capabilities allow the code to be applied to complex geometries. Commercial CFD codes such as FLUENT are widely used in many industries today and are commonly used to predict mixing phenomena.

Each of the steps in a CFD analysis can influence the predicted results and should be considered. The basic steps are describing the physical model; defining the CFD model domain, boundary conditions, and models; validating the solution; and completing sensitivity studies. When considering CFD predictions, the analyst must consider the assumptions and limitations of each step in the process. Further details on the fundamentals of CFD are found in the introductory text by Anderson.[5]

2 OVERVIEW OF ANALYSIS

The analysis is carried out in a series of steps that focus on specific issues affecting the mixing behavior. Each prediction is completed using modeling assumptions consistent with the successful validation analysis completed at 1/7th scale (Ref. 4). Similar to the 1/7th scale predictions, a quasi-steady assumption is made. The boundary conditions are selected to represent a fixed time during the transient. Transient simulations are not practical due to the relative size of the steam generator models and the extended time span of the transient. The analyses completed are outlined below.

2.1 Scaleup

The first sets of analyses are focused directly on the issue of scaling. These preliminary analyses are considered necessary to separate the effects of scale and geometry. Ultimately, a comparison of the 1/7 h scale predictions and predictions for a prototypical Westinghouse steam generator under severe accident conditions is desired. But the 1/7th scale facility utilizes steam generators that are not completely similar to the prototypical Westinghouse design. The facility data could contain both geometric and scale distortions when compared to the prototypical Westinghouse analysis. To isolate the effect of scale, this set of predictions is completed using a full-scale hot leg and steam generator that are geometrically similar to the 1/7th scale test facility. Two sets of full-scale boundary conditions, obtained from a SCDAP/RELAP5 analysis of the ZION nuclear power plant under severe accident conditions, are used. The two sets of predictions are compared to the 1/7th scale results to quantify the effect of scale on key mixing parameters. In addition, a sensitivity study based on the secondary side heat transfer rate is completed to quantify the effect of this significant governing parameter.

2.2 Prototypical Westinghouse Steam Generator

A second set of predictions is completed for a prototypic Westinghouse model 44 steam generator. These predictions are similar to the scaleup predictions completed earlier except for the geometry. All model and boundary conditions remain consistent. These predictions, compared with the scaleup results, provide a direct indication of the effect of the geometric differences on the inlet plenum mixing parameters. A further comparison of these predictions with the 1/7th scale data provides an indication of the applicability of the 1/7th scale data for representing the behavior of a prototypic Westinghouse steam generator. It is noted that the primary side dimensions of a model 44 steam generator are very similar to the dimensions of the model 51 designs.

2.3 Effect of Tube Leakage

The third set of predictions focuses on the effect of a leaking tube or tubes and how this mass loss affects the inlet plenum mixing and entrainment. Steam generator tubes may contain small leaks that could grow during this type of severe accident scenario. If the leak adversely affects the inlet plenum mixing or draws hot gas directly to the leak, the leaking tube could fail earlier than expected. The 1/7th scale tests do not address this issue. Boundary conditions and geometry, with the exception of the leak, are consistent with the scaleup predictions to provide a direct comparison between predictions with and without a leak.

3

2.4 Prototypical Combustion Engineering Plant Steam Generator

The fourth set of predictions is based upon a prototypic steam generator design from a Combustion Engineering (CE) plant. The CE inlet plenum geometry is significantly different than in the Westinghouse design. Mixing is expected to be less effective due to a reduced distance from the hot leg nozzle to the tube sheet entrance. Predictions are completed in temperature ranges similar to the Westinghouse analysis. The boundary conditions are obtained from SCDAP/RELAP5 predictions of the Calvert Cliffs nuclear power plant conditions during a severe accident. These predictions give an indication of the expected inlet plenum mixing for this geometrically different steam generator design.

2.5 Steam Generator Geometry

Each of the steam generator models consists of one hot leg and steam generator primary side. The model begins with the hot leg at or near the vessel exit and ends at the outlet plenum exit of the steam generator. Each hot leg section is straight leading up to the elbow and inlet plenum nozzle. During the proposed transient, the loop seals are filled with water. The outlet plenum nozzle and piping are therefore not included.

Details of the steam generator models are beyond the scope of this report. Each model is constructed from copies of original drawings construction drawings. Some minor dimensional changes are made to facilitate the mesh generation. The models used are a good representation of a prototypical steam generator. Figure 2 shows the primary flow paths of a generic steam generator model with some overall dimensions to provide an indication of the basic steam generator size and hot leg orientation. Table 1 lists the dimensions of each of the steam generators modeled. For each model, only the primary flow path is considered. Boundary conditions are applied at the inner wall to eliminate the need to model the wall material and thickness.

The steam generator tubes are modeled using an approach developed and used successfully at 1/7th scale (Ref. 4). Some parameters of the tube bundle geometry are given in Table 1. A coarse tube model is developed to have the prototypical heat loss and pressure drop characteristics. Tubes are grouped together in this approach to form a reduced number of individual tube flow paths. Details of the tube modeling approach are addressed later in this report.

3 CFD MODEL DEVELOPMENT

The physical geometries described above are represented with a finite volume mesh on which the governing Navier-Stokes equations are discretized and solved. All models are developed for the FLUENT version 6.0 unstructured CFD code.

To save computer resources, a symmetry assumption is applied where possible. Symmetry is applied to the scaleup model and the CE plant steam generator. The Westinghouse model 44 steam generator is not symmetric. A transient solver is applied with steady boundary conditions to obtain a solution for each model. Many solutions resulted in some oscillatory plume behavior that required the transient solver approach in order to obtain a converged solution. The principal features used for each CFD model are summarized below.

- ▸   transient Reynolds-averaged Navier-Stokes solution with steady boundary conditions
- ▸   vertical symmetry plane (for symmetric designs)
- ▸   Reynolds stress turbulence model (2nd order) with nonequilibrium wall functions
- ▸   full buoyancy effects on turbulence (as defined in FLUENT model)
- ▸   temperature-dependent thermal properties (steam) at constant pressure (2400 psia)
- ▸   gravity
- ▸   segregated solver with 2nd order differencing on momentum and energy
- ▸   porous media model representing steam generator tube flow paths

Specific features of the models and major assumptions are outlined below.

3.1 Finite Volume Mesh

Geometric details of the hot leg, nozzle, and steam generators are used to develop a mesh suitable for use with the FLUENT CFD code. An effort is made to produce models with a high-quality mesh that represents the key features of the primary system flow path for the hot leg and steam generator. To maintain a level of consistency between models, similar mesh quality and spacing are used for each. In the bulk of the mesh, the cell aspect ratios are limited to 2 with the exception of the tube bundles, where cell aspect ratios are stretched to 5 in the flow direction. Cell skew is minimized through careful grid spacing and with the use of hexagonal elements. Growth rates between neighboring cells are limited to 20% with few exceptions. The majority of the cells in the hot leg and plenums are of the same approximate dimensions.

Some mesh characteristics for the individual models are given in Table 2. The number of computational cells refers to a full model (without symmetry applied). The models utilizing the symmetry assumption used half of the number of cells reported. The number of cells for a full model is listed in the table to simplify the comparison between models. The CE plant steam generator is significantly larger and required a larger number of cells as indicated in the table. The final row in Table 2 lists the average cell dimension. This value is the cube root of the average cell volume. Cell dimensions in the inlet plenums and near walls are typically less than half of this average value and cells in the tube bundle are larger.

Figure 3 shows the mesh for the scaleup model with the mesh mirrored across the symmetry plane to provide a full steam generator view. This mesh is identical to the mesh used for the 1/7th scale work reported earlier (Ref. 4). The tube leakage predictions also utilize this mesh. Figure 4 shows the mesh used to model the Westinghouse model 44 steam generator. This is

5

the only mesh not to incorporate a symmetry plane. The model 44 mesh characteristics are generally consistent with the scaleup mesh. Figure 5 shows the mesh used to model the CE plant steam generator. This figure is also mirrored across the symmetry plane to provide a full view. This geometry is significantly larger than the Westinghouse model 44 designs.

3.2 Boundary Conditions

Boundary conditions are varied between models and for the sensitivity studies. However, many common attributes apply to all models. Each model utilizes a no-slip boundary condition for the hot leg and inlet plenum walls. Tubes are modeled with a porous media approach that is described later. In addition, the hot leg and inlet plenum walls are assumed adiabatic. For the scaleup and CE plant models, the vertical mid plane of the steam generator and hot leg is modeled as a symmetry plane to reduce the size of the model by 50%. The vessel end of the hot leg is split into two boundaries to facilitate an inflow and an outflow condition. A 50/50 (60/40 for the scaleup model) split based on height is used for the end of the pipe. Small variations in this ratio are found to have little or no effect on the results. Hot gas enters the upper portion of the hot leg at the vessel end at a rate and temperature obtained from SCDAP/RELAP5 predictions of an appropriate nuclear power plant during a severe accident. The only heat transfer from the system occurs in the tube bundle above the tube sheet. The tube sheet portion of the tube bundle model is assumed to be adiabatic. The heat transfer from the tube bundle is adjusted using the tube bundle model to be consistent with the heat transfer predicted from SCDAP/RELAP5 analysis. Uncertainty in this important boundary condition is addressed by considering a wide range of heat transfer rates in a series of sensitivity analyses. Figure 6 gives an overview of the boundary conditions used for the steam generator models. Specific boundary conditions and assumptions are outlined below.

3.2.1 Tube Modeling

The tube modeling approach was successfully used in previous work at $1/7^h$ scale (Ref. 4). A similar approach is used for these full-scale models. The resulting tube model reproduces the overall flow and heat transfer characteristics of the prototypical tube bundle with a relatively small number of computational cells.

The previous work at $1/7^{th}$ scale utilized a single flow path for each of the 216 tubes in the test facility. This one to one approach is not practical for a prototypical steam generator with thousands of tubes. To reduce the mesh requirements, neighboring tubes are grouped together to form a reduced number of flow paths. Each flow path used in the CFD model is established to reproduce the pressure drop and heat transfer characteristics of the group of tubes it represents. Figure 7 shows a grouping of tubes and the corresponding flow channel used in the tube bundle model. This approach does not resolve flows in an individual tube but does provide a reasonable upper boundary condition for the inlet (and outlet) plenum where the mixing of interest takes place. The principal attributes of the tube bundle model are the pressure drop at the tube entrance, the flow (viscous) losses within the tubes, the heat losses from the tubes, and the buoyancy driving forces. Tube flows are determined by a balance between the flow losses (viscous and inertial) and the buoyancy driving force. Results at $1/7^{th}$ scale suggest this technique is a good approach.

The pressure drop and viscous loss coefficients determined for the simplified tube bundle model are obtained from detailed CFD predictions of flow losses from a small group of prototypical tubes. The tubes are modeled in great detail with models containing up to 1 million cells to ensure the details of the flow boundary layers and tube entrance losses are captured. From the predicted values of flow loss over a wide range of flow and temperature conditions, loss coefficients are determined for a simplified single channel porous model of the group of tubes. Tube entrance flow gradients and tube wall shear layers are not resolved in the simplified model. The loss coefficients provide the appropriate flow entrance pressure drops and shear losses along the tube. In this way, thousands of tubes are grouped together to form hundreds of individual flow paths. This approach allows the model to predict the tube bundle flow characteristics with a practical number of computational cells. The heat transfer characteristics of the simplified tube model are also adjusted to achieve the expected rate of heat loss from the tube bundle. An example of the tube bundle modeling approach for a single tube is given in Appendix A of Reference 4. The approach used here repeats that process using multiple tubes as shown in Figure 7

3.2.2 Velocity Inlet Conditions

The velocity inlet covers a region that represents the upper 60% (scaleup model) or 50% (model 44 and CE model) of the height of the vessel end of the hot leg (Fig. 6). It is assumed that the flow coming from the reactor vessel is well mixed and at a uniform temperature. A uniform velocity is also defined at the vessel end of the hot leg. Sensitivity studies at 1/7[h] scale indicated that flows at the steam generator end of the hot leg are not significantly affected by changes in the inlet velocity profile at the vessel end of the hot leg. The uniform velocity profile is utilized for its simplicity and for lack of a defendable alternative. The temperature at the velocity inlet is adjusted to match the predicted hot leg temperatures from SCDAP/RELAP5 transient simulations. The magnitude of the inlet velocity is adjusted to obtain the desired mass flow in the hot leg. The result is a hot leg mass flow and temperature that are consistent with the SCDAP/RELAP5 predictions.

3.3 Material Properties

The working fluid used for each model is steam at 2400 psia. Fluid properties are obtained from the system code database used by the SCDAP/RELAP5 models for steam. Specific data covering the range of expected temperatures are input to the FLUENT code in tabular form. Since the quasi-steady analysis is conducted at a fixed pressure, the properties are specified as temperature dependent only. The FLUENT code uses linear interpolation (based on temperature) to find the thermal properties from the table.

3.4 Turbulence Modeling

The second order Reynolds Stress Model (RSM) is utilized for all predictions. Detailed experimental data, which are not available at these conditions, are needed to fully validate the selection of a turbulence model. The second order RSM turbulence model provides excellent predictions at 1/7th scale and is considered to be the most appropriate turbulence model available in the FLUENT code for this type of flow pattern. The RSM model does not assume isotropic turbulence like common two-equation k-epsilon type models. Full buoyancy effects (as defined in the FLUENT code) on the turbulence are applied and nonequilibrium wall functions are used. Analysis at 1/7th scale with several turbulence models demonstrated that the turbulence model selection did not significantly alter the overall solution behavior. Further details of the turbulence models are beyond the scope of this report.

3.5 Solution Convergence

Steady-state solutions are obtained by using a transient solver with steady boundary conditions. Initially, each case is run using the steady-state solver to obtain a rough solution that serves as the initial condition for the transient solver. Oscillations in the solutions, such as the buoyant plume oscillations, appear to prevent the steady-state solver from complete convergence. The transient solver is applied until a steady solution is obtained. Solution convergence is monitored in several ways. First, solution residuals are monitored at each time step to ensure a sufficient reduction is achieved. In addition, several temperatures and velocities are monitored at key points to verify the solution has reached a steady state (or steady oscillatory state). Finally, the overall mass and energy balance are monitored for convergence.

3.6 Grid Independence

A complete grid independence study is not performed due to the physical scale of the model. A high-quality mesh is used for this analysis to minimize any grid effects on the results. Second order differencing is used to reduce numerical diffusion. In addition, the mesh quality remains consistent from model to model to facilitate the comparison of results from different designs. Qualitative information on grid independence is given below for completeness.

The mesh used for each of these models is built with an experience gained from the many grids used to develop the 1/7th scale model (Ref. 4). Each mesh is a compilation of the lessons learned during this process. Grid optimizations include optimized node density for wall functions and verification of grid independence in the tube bank. The final models use hexagonal elements with grid stretching and cell skew minimized. In regions of high gradients or transitions, the cell aspect ratio is kept very close to 1. Mesh size is reduced at the walls to accommodate the wall functions. Transitions away from the wall are limited to growth rates between 5% and 20%.

Although a full grid independence study is not completed, a high degree of confidence in the quality of the mesh is obtained from localized grid studies, the high quality of the mesh, and a good prediction of the experimental data at 1/7th scale using a similar mesh.

4 SUMMARY OF ASSUMPTIONS AND LIMITATIONS

The completed CFD predictions give valuable insights into the three-dimensional mixing and entrainment phenomena in a steam generator inlet plenum. However, all predictions are affected by the modeling approach and the various assumptions that are made. In some instances, more work may be needed to go beyond the initial evaluations documented in this report. To facilitate a clear understanding of the predicted results and to highlight areas of potential further study, some major assumptions and limitations of the modeling approach are provided below.

4.1 Tube Bundle Model

The tube bundle model is designed to match the pressure drops and heat transfer characteristics of the actual tube bundle. The goal is to preserve tube bundle mass flows by correctly modeling the buoyancy driving force and the pressure loss terms without modeling the individual tube flows in detail. Appendix A from Reference 4 provides the details of how the tube bundle model is developed.

The tube model results in tubes with a flow cross-sectional area that is larger than the prototypical section of tubes it represents (see Fig. 7). With mass flow and temperature preserved, the tube flow velocity is significantly lower in the tube bundle model. A lower velocity in the tube bundle affects the tube flow residence time. The time for the flow to pass through the tubes is not considered significant in light of the type of steady-state analysis that is being completed. The frequency or magnitude of solution oscillations, however, could be affected by the tube flow residence time.

Tube flows returning to the inlet plenum are another aspect of the solution where the tube bundle model impacts the solution. In the prototype, the flow returns through the tubes at a higher velocity than the tube bundle model (due to the change in cross-sectional area). More than one thousand individual jets of return flow enter the top of the inlet plenum and impact the local turbulence and mixing in the prototype. The predictions provide the same mass flow entering the inlet plenum at a lower velocity. The predictions indicate that flows along the tube sheet face are directed away from the hot plume. Return flows are therefore swept away from the hot plume. The tiny jets emanating from the tube sheet are expected to dissipate quickly and are not expected to impact the global inlet plenum mixing.

An obvious limitation of the tube bundle model is the reduced number of tubes. Thousands of tubes are grouped and modeled with hundreds of individual flow paths. This simplification limits the resolution of tube-tube variations and other variables across the tube bundle. The resolution obtained, however, does provide temperature and velocity profiles across the tube bundle and is considered adequate for the purposes of this analysis.

Finally, the tube bundle heat transfer is greatly simplified. It has been demonstrated that the tube bundle heat transfer does affect the tube bundle mass flows and other parameters important to the mixing. Simplifications in this area could be significant. Secondary side conditions that lead to variations in the tube wall boundary conditions have not been investigated here. All tubes (at all locations) are subjected to the same external heat transfer coefficient and sink temperature. There is no tube bundle structural mass in this steady state analysis. The tube walls are infinitely thin. Ranges of heat transfer rates have been considered in this analysis to cover the range of all potential heat transfer from the tube flows.

4.2 Grid Independent Solution

A rigorous grid sensitivity study is not completed due to the large scale of the computational models. Reducing the cell dimensions by 50% in a 1 million cell model results in 8 million computational cells. This size model is not practical.

Grid independence is addressed in a variety of ways to try to verify that the grid does not affect the predictions. Local grid refinements are completed to maintain adherence to optimal grid size characteristics. For instance, wall cell sizes are adjusted to provide the correct range of sizes for the wall functions. In addition, a careful analysis is completed to verify that the tube bundle mesh is adequate to resolve the flow loss terms. Careful attention is paid to the grid quality and consistency from model to model to reduce grid dependence between the various models. Cell growth rates are limited to 20% or less with few exceptions. Cell skew is minimized through careful node spacing on edges and boundary layers. The grid dependence issue is not addressed directly but care is taken to ensure a high-quality mesh that is consistent from model to model.

4.3 Adiabatic Walls

The steam generator models lose heat from the primary system only through the tube bundle walls. The hot leg and inlet plenum walls are assumed adiabatic. In steady-state analysis, the model walls are not heating up and removing heat from the system. Fixed heat transfer rates could be specified but the magnitude and spatial variations of these values would add uncertainty to the calculations. The approach used simplifies the models with the adiabatic assumption. This provides a consistent boundary condition across each of the steam generator models and is also consistent with the 1/7th scale predictions completed earlier (Ref. 4).

A quick review of SCDAP/RELAP5 results from the ZION case used for the high-temperature model 44 predictions indicates that over 81% of the heat leaving the hot leg/steam generator system goes into the tube bundle. The system code analysis indicates that approximately 7% of the heat is going into the hot leg wall and the remaining heat is going into the walls of the inlet and outlet plenum (including the tube support sheet and the divider plate). Any potential impact of this heat loss on the inlet plenum mixing is not addressed.

4.4 Symmetry Model

A symmetry plane is used for the full-scale scaleup and the CE plant steam generator models. This symmetry assumption is expected to have an impact on the rising hot plume's oscillatory behavior. The model 44 predictions demonstrated significantly increased plume oscillations when compared to the nearly equivalent full-scale simulations or the CE plant model. The model 44 inlet plenum is not symmetric and no symmetry plane was assumed. It is not clear whether the actual asymmetry of the model 44 design enhances the oscillatory behavior of the plume or whether the symmetry plane used in the other models diminishes this effect. Simulations without a symmetry plane assumption were not completed for the symmetric scaleup model or the CE plant model.

4.5 No-Radiation Model

The temperatures reached during the period of rapid core oxidation in this event are high enough to generate thermal radiation transfer processes between the hot gas and the system walls or other cooler gas flows. Thermal radiation is not accounted for in these analyses. Radiation is expected to transfer heat from the hot gas flows in the hot leg to the cooler gas returning to the reactor vessel. Radiation transfer of this type is expected to reduce the temperature of the flows reaching the tube bundle. Radiation is also expected in the inlet plenum region and the tube flows in the entrance region of the tubes.

4.6 Fixed Boundary Conditions

The boundary conditions are all fixed during these simulations. No feedback to the system is possible. The hot leg mass flow is taken from the single SCDAP/RELAP5 prediction. Since the vessel is not modeled, the source of flow to the hot leg is fixed and does not respond to changes in the system. For instance, the rate of heat loss from the tube bundle is expected to influence the amount of flow from the vessel to the steam generator. This effect is not considered. Another area of concern relates to the tube leakage analysis. Significant leakage at the steam generator tube bundle is expected to draw additional flow from the reactor vessel towards the leaking steam generator. This effect is not accounted for. Mass flows from the reactor vessel remain unchanged for all cases where tube leakage is applied.

A more general issue relates to the validity of the SCDAP/RELAP5 predictions used for boundary conditions. The flow from the vessel that enters the hot leg is determined from the SCDAP/RELAP5 predictions. It is recognized that the natural circulation flows in the vessel, hot leg, and steam generator are coupled together. Changing the flow conditions for one of these regions is expected to impact the flows in the other regions. This balancing of the overall flows is completed in SCDAP/RELAP5. The uncertainty in these predictions is unknown.

4.7 No Hydrogen

During the period of rapid core oxidation, significant hydrogen is expected to build up in the reactor system. This hydrogen can affect the heat transfer rates and flow rates in the hot leg and steam generator tubes. A hydrogen bubble could potentially plug some of the tubes in the bundle. The high temperature predictions will be most impacted by this assumption. These effects are not considered in this analysis.

4.8 Quasi-Steady Solution

Fixed boundary conditions are applied to obtain a steady-state solution that represents a snapshot in time of the transient behavior. The fixed boundary conditions are obtained from a single point in time selected from SCDAP/RELAP5 transient predictions. Experimental results at $1/7^{th}$ scale indicated that the quasi-steady assumption is reasonable.

One area where this assumption can have an effect is on the tube flows that are influenced significantly by the tube heat transfer rate. Tube flow residence time is approximately 1 minute based on the model 44 steam generator predictions. During this time, the hot leg inlet temperature can change by over 100 degrees Kelvin during the period of peak temperature rise.

11

This implies that flow returning from the tubes had originally entered the tubes at a temperature 100 degrees lower (estimate) than the current tube flow temperatures. The boundary conditions used for the steady-state solution are a snapshot in time from the SCDAP/RELAP5 predictions and implicitly take this feature into account. However, there is some uncertainty in the ability to fully account for this transient effect. General uncertainty and other simplifications used to model the tube bundle heat transfer are expected to be more significant than this quasi-steady assumption.

Another issue not addressed by the quasi-steady assumption is the relief valve cycling. All steam generator solutions are based on boundary conditions selected from time periods where the relief valves are closed. Relief valve cycling is not addressed. The experimental evidence at 1/7th scale suggests that the flow pattern reestablishes itself very quickly after the relief valve closes. No attempt is made here to study the effect of the relief valve cycling.

5 SCALEUP ANALYSES

The first set of full-scale predictions is completed to isolate the effect of scale and facilitate a consistent comparison with the previously reported (Ref. 4) 1/7th scale predictions. Potential effects of geometrical distortion between the 1/7th scale facility and full-scale geometries are eliminated. The geometry and mesh used at 1/7th scale are scaled directly to full-scale. All cell dimensions are simply increased by a factor of 7, resulting in a model with complete geometric similarity to the 1/7th scale model.

Prior analysis at 1/7th scale indicates that the secondary side heat transfer rate is a dominant governing parameter. During this analysis, a range of secondary side heat transfer rates is used to quantify the significance of this parameter at full scale. The range of heat transfer rates is selected to ensure that the heat transfer rates cover a range similar to both the 1/7[h] scale tests and the expected prototype conditions. Direct comparisons between models of different scale are made when the heat transfer rates are consistent between the models. This consistency is measured in terms of the normalized rate of decrease of the tube temperatures for the tube flows.

Temperatures are made dimensionless using the following relation:

$$T' = (T - T_{ct}) / (T_h - T_{ct})$$

The tube return flow temperature, T_{ct}, and the hot flow temperature in the hot leg, T_h, are used as the low- and high-point reference temperatures. These represent the limits of low and high temperatures entering the inlet plenum, which serves as a mixing chamber. Hot leg flows enter the inlet plenum at a dimensionless temperature near 1 and cold tube return flows enter the inlet plenum at a dimensionless temperature near 0. Dimensionless tube entrance temperatures near 1 indicate a complete lack of mixing. Similarly, dimensionless tube entrance temperatures closer to 0 indicate a significant amount of inlet plenum mixing.

5.1 Scaleup Boundary Conditions

Two sets of analyses are completed using boundary conditions from SCDAP/RELAP5 predictions of the transient behavior for the ZION nuclear power plant during a TMLB' station blackout event. Some variables from these predictions are plotted in Figure 8 to give an indication of the transient progression. Two times from the SCDAP/RELAP5 predictions are selected for a quasi-steady CFD analysis. At each of the times, a series of CFD predictions are made over a wide range of secondary side heat transfer rates. The times used for the analysis are noted on Figure 8. The high temperature h case is completed when the hot leg flow temperature is 1444 K. This point in time is near the time predicted for the primary system boundary failures. A second set of predictions is made using conditions from a point earlier in the transient when the hot leg temperatures are 1024 K. This condition is referred to as the low-temperature l case. These two points cover a wide range of the expected conditions. The boundary conditions predicted by SCDAP/RELAP5 that are used for the high- and low-temperature cases are listed in Table 3. Mass flows in the table represent the predicted plant conditions and do not account for symmetry in the CFD model.

The flow inlet boundary condition is established at the vessel end of the hot leg. Hot steam flows into the model on the upper 60% of the pipe (height) with a uniform temperature and velocity. The temperature value for each condition is listed in Table 3 and a velocity is established to provide the associated mass flow rate. Predictions reported later in this report refer to hot leg mass flows and temperatures determined on a vertical plane in the hot leg midway between the vessel and steam generator. These results differ slightly from the applied boundary conditions due to mixing and entrainment in the hot leg. This point is made here to avoid confusion later.

The secondary side temperature (Table 3) used for heat transfer from the steam generator tubes is a representative value obtained from the SCDAP/RELAP5 predictions. A uniform value for the heat transfer coefficient and secondary side temperature is applied to the external wall of all the tubes. Tube heat transfer is also controlled with an effective tube bundle thermal conductivity value that is established as part of the porous media assumption used for the tube model (see Appendix A, Ref. 4 for a description of the tube bundle modeling approach). A set of sensitivity studies is completed on the secondary side heat transfer rate. This rate is varied to cover a range of conditions that spans the expected prototype behavior and the 1/7th scale test data. Sensitivity studies for the high-temperature case are labeled h1 (lowest heat transfer rate) through h7 (highest). The low-temperature cases are labeled l1 through l7. The absolute value of the heat transfer coefficient used is not directly comparable to the prototypic steam generator tube value due to the differences in the heat transfer and flow characteristics between the prototype tube bundle and the porous media tube model. Heat transfer rates are compared by looking at the rate of temperature drop in the tube flows. As the steam flows through the tube bundle, the temperature of the flow asymptotically approaches the secondary side temperature. The rate of tube bundle heat transfer affects the rate of this temperature change.

One purpose of the scaleup simulations is to quantify the effect of scale on the flow and mixing behavior in the hot leg and steam generator during the type of natural circulation flow illustrated in Figure 1. This is accomplished by comparing the full-scale predictions with 1/7th scale predictions completed earlier on a similar geometry. Key scaling parameters are given in Table 4 to compare the 1/7th scale test documented in Reference 4 and the present analysis at full-scale conditions. A Reynolds number is computed for the hot leg flow using the hot leg radius for a length scale. The hot flow conditions are used for the density, viscosity, and average velocity. The full-scale Reynolds number is roughly a factor of 3 higher than the Reynolds number achieved in the 1/7th scale tests. A Grashof number, indicating the ratio of buoyancy forces to viscous forces, is computed for the hot plume rising through the inlet plenum. The inlet plenum radius is used for a length scale. The full-scale Grashof number is roughly one order of magnitude higher than the value obtained at 1/7th scale. A Richardson number, computed as the ratio of Gr/Re^2 using the values with different length scales from above, provides an indication of the relative importance of free (Ri>>1) and forced convection (Ri<<1). The Richardson numbers computed are nearly the same for the full-scale and 1/7th scale predictions. Generally speaking, the flows from the 1/7th scale test result in scaled parameters within an order of magnitude of the full-scale predictions. From a fluid dynamics perspective, the 1/7th scale tests are reasonably scaled to represent full-scale behavior. Issues such as radiation, geometry variation, and other prototypical phenomena are not addressed here.

5.2 High Temperature Scaleup Predictions

Qualitatively, these full-scale predictions are very similar to the 1/7th scale predictions documented in Reference 4. The flow patterns observed at 1/7th scale and documented in Reference 4 are qualitatively the same as those observed under these full-scale conditions.

Figure 9 illustrates the mass-averaged dimensionless temperature profile along the flow path of the hot tubes as a function of a dimensionless height for the high-temperature case. A length scale representing the average distance from the tube entrance to the top center of the tubes is used to nondimensionalize the height. A height of zero represents the tube sheet entrance and a height of 1.0 indicates the highest point of the U bend in the tubes. The hot tubes are tubes that carry flow from the inlet plenum to the outlet plenum.

Figure 9 clearly shows the effect of changing the heat transfer rate to the secondary side. The highest heat transfer rate applied (case h7) reduces the bulk average hot flow temperature to the secondary side temperature at a normalized height of approximately 0.35. Case h1, with the lowest rate of heat transfer, predicts temperatures that are still significantly higher than the secondary side temperature at the top of the tube bundle.

Case h5 utilizes a heat transfer rate that results in the mass-averaged hot tube temperatures reaching the secondary side temperature over a path length nearly equal to the tube bundle height. This heat transfer rate is consistent with the experimental observations from the 1/7th scale test examined earlier (Ref. 4). The 1/7th scale predictions are shown on Figure 9 as the hexagons. Case h5 is comparable to the 1/7th scale results in terms of consistent heat transfer rates and is subsequently used for direct comparisons to the 1/7th scale predictions.

Predicted tube temperatures for the hot case from the SCDAP/RELAP5 analysis are illustrated on Figure 9 as the four solid-line segments. These segments represent the data from the four upward flow hot tube bundle cells in the SCDAP/RELAP5 model. The span of the horizontal line segments is equal to the vertical length the individual cells. The results indicate a temperature reduction rate most similar to case h1. One notable difference is that the system code results start at a lower temperature than the CFD predictions. This is explained by the differences in the heat transfer assumptions between the two predictions. The CFD model applied an adiabatic assumption for all surfaces except for the tubes above the tube sheet. The flows in the SCDAP/RELAP5 model lose heat to the inlet plenum walls and the tube sheet, which accounts for the lower initial temperature on Figure 9. The boundary conditions of the CFD model match the hot leg temperature to the SCDAP/RELAP5 predictions.

The tube outer wall heat transfer coefficient used in the SCDAP/RELAP5 predictions shown on Figure 9 is based on a Nusselt number of 4 resulting from a laminar flow assumption. The external heat transfer from the tubes is essentially conduction through a gas. Since some natural circulation is expected on the secondary side of the tube bundle, the actual rate of heat transfer is expected to be higher than the value applied in the SCDAP/RELAP5 model. Radiation could also enhance the secondary side heat transfer. The SCDAP/RELAP5 predictions used are assumed to represent the low end of the expected heat transfer rate from the tubes.

Table 5 lists some of the key parameters determined from the high-temperature predictions over the range of heat transfer rates (h1-h7). In addition, the predictions at 1/7th scale are included to provide a direct comparison with case h5. Figure 10 plots selected variables from the table as a function of increasing heat transfer rate to help identify the trends in the table.

The first row in Table 5 lists the total heat loss from the tube bundle. The tabulated tube heat loss is obtained as a surface integral of the heat flux over the total tube surface and is predicted to decrease as the heat transfer rate is increased. These predictions seem contradictory but can be explained by looking at some of the other predicted results. Tube heat loss is proportional to the product of the temperature drop in the tubes ($T_{ht} - T_{ct}$) and the tube bundle mass flow rate (m_t) from a first law of thermodynamics analysis. Although the temperature drop in the tubes increases with increasing tube heat transfer, there is a significantly larger decrease in the tube bundle mass flow over the same range of heat transfer rates. Taken together, changes in the temperature drop and the tube bundle mass flow account for the decrease in the integrated tube bundle heat loss with increasing heat transfer rates.

The percentage of tubes carrying hot gas is given in the second row of table 5. This value is also plotted in Figure 10. The number of hot tubes reaches a minimum near cases h4 and h5. It is suggested that this minimum in the number of hot flow tubes is related to a maximum in the buoyancy driving force expected to occur near cases h4 and h5. These cases correspond to heat transfer rates that effectively reduce the hot tube flow temperature to the secondary side temperature as the flow reaches the top of the tube bundle (see Fig. 9). Lower heat transfer rates result in increased temperatures in the downflow (outlet plenum side) tubes and would tend to reduce the buoyancy force. Higher heat transfer rates reduce the height of the hot flow column in the upflow tubes (inlet plenum side), also reducing the buoyancy driving force. Therefore, the conditions associated with cases h4 and h5 are expected to create the largest "chimney" effect in the tubes. As the rising hot inlet plenum plume impacts the tube sheet it partially stagnates on the surface (spreading radially outward) and partially penetrates into the tubes. A large "chimney" (buoyancy) effect is expected to reduce the portion of the plume that stagnates and allow the flow to penetrate into the tubes with a smaller amount of spreading. This trend is consistent with the data for cases h4 through h7. At the highest heat transfer rate, h7, the hot tube flow quickly reaches the secondary side temperature and the bundle mass flow is the lowest value predicted in this study. The rising hot plume stagnates (and spreads radially) more in this case and the larger percentage of hot tubes for case h7 is consistent with this. For lower rates of secondary side heat transfer the situation appears more complex. Although the percentage of hot tubes does increase as the heat transfer rate is reduced from case h5 to case h1, the overall tube bundle mass flow increases over this range.

The percentage of hot tubes from case h5 is compared directly to the 1/7th scale predictions. Case h5 is selected for this comparison since it exhibits a similar normalized rate of temperature drop in the tubes. These two cases are in complete agreement (38% of the tubes) for the number of hot flow tubes. These results suggest that the 1/7th scale test data for the number of hot flow tubes does represent full-scale behavior for similar geometry and a consistent heat transfer rate. Figure 11 shows the predicted dividing line between the hot and cold flow tubes at the tube sheet face for case h5 and the 1/7th scale results. Even the layout of the hot and cold tubes for cases h5 and the 1/7th scale prediction is similar. They differ by one group of tubes. Case h1 is also illustrated in Figure 11. The boundary for case h1 shows the extent of the spreading of the hot tube region beyond the results for h5. An additional 18 tube sections (8.3% of the bundle) carry hot flow in case h1 compared to case h5.

The hot leg hot-flow temperature, T_h, and the hot leg mass flow, m, are not significantly affected by the changing heat transfer rate in the tubes. This results from the fixed boundary condition assumption at the vessel end of the hot leg. Integral primary system behavior, specifically the hot leg mass flow, is assumed to be sensitive to changes in the heat transfer rate from the tube bundle. This aspect of the system behavior is not addressed in this model since a fixed mass flow and temperature are applied at the hot leg (vessel) end. The temperature of the cold flow in

16

the hot leg is affected by the heat transfer rate as shown in the table. The lowest heat transfer rates result in the lowest temperatures returning from the inlet plenum. This is consistent with the increased tube heat loss and the increased mixing predicted for the cases with the lower tube heat transfer rates.

The temperature of the flow entering the hot tubes is plotted on Figure 10 as a function of the heat transfer rate. This value is lowest for the minimum heat transfer rate. As suggested earlier, this is attributed to an increase in the inlet plenum mixing due to the significantly increased tube bundle mass flows at the lower heat transfer rates. Although the temperature of the flow entering the tubes is lower for lower heat transfer rates, the total challenge to the tubes may not be. The lower temperature values also correspond to significantly higher mass flow rates and significantly higher total energy transfer to the tubes. A detailed structural analysis of the tube bundle is needed to quantify the challenge to the tubes for each case.

The ratio of the tube bundle flow to the hot leg flow is referred to as the recirculation ratio. Since the hot leg mass flow changes very little, the recirculation ratio shows the same general trend as the tube bundle flows. Both the tube bundle mass flow and the recirculation ratio are plotted on Figure 10 as a function of the heat transfer rate. Recirculation ratios from 1.85 to nearly 3 are predicted. Clearly, this value is impacted significantly by the tube bundle heat transfer rate. A comparison of the recirculation ratio for case h5 and the 1/7th scale predictions shows good agreement (less than 5% difference). These results suggest that the 1/7th scale data provide a good representation of the full-scale recirculation ratio for consistent heat transfer rates (and a consistent geometry). The tube bundle flows result from the hot flow entering the tube bundle and rising up into the tubes. A large recirculation ratio is an indication of significant entrainment of the cooler inlet plenum steam into the rising hot plume as it passes through the inlet plenum and enters the tubes.

The mixing fraction is a parameter used to define the portion of the hot leg hot flow that passes directly into the hot tubes without mixing. Based upon a simple mixing model, a mixing fraction of 0.85 implies that 85% of the hot leg hot flow is mixed completely in the inlet plenum with flows returning from the cold tubes. The remaining hot leg flow, 15%, goes directly into the tubes (where it completely mixes with a much larger flow from the inlet plenum at the mixed mean temperature). The predicted mixing fraction, listed in the last column of Table 5, does not change significantly with the tube bundle heat transfer rate. Mixing fraction is plotted on Figure 10. It is unclear if the elevated values for cases h5 and h6 are significant. A comparison of the mixing fraction with the 1/7th scale prediction shows a small increase at full-scale conditions. The mixing fraction for case h5 is almost 14% higher than the 1/7[h] scale prediction. It is unclear if this variation is significant in light of the variations observed over the complete set of tests at 1/7th scale and the sensitivity of this parameter to small changes in temperature.

Tube flow entrance temperatures are a primary result from this analysis. In order to study the challenge posed to the tube bundle, the tube flow temperature and mass flow are needed. The flow temperatures further into the tubes are not reported here since they are affected somewhat by the assumptions and limitations of the tube bundle modeling approach. Tube entrance temperatures, however, are less affected by the tube model and are suitable as reference values for comparisons between predictions. For detailed tube temperature data at various locations throughout the tube bundle, a full transient thermal analysis is needed that incorporates the tube sheet and tube wall metal mass. This is not part of the present model.

Predicted tube entrance temperatures for case h5 are shown in Figure 12 as a horizontal contour plot located 7 inches (0.1778m) above the tube sheet entrance. The 7-inch level is selected to correspond to the 1-inch (0.0254m) level used to report the 1/7th scale results. Variations in the temperature are observed between the tube sections (tube-tube) and also within some of the individual tube sections. Variations within each section are highest near the tube sheet entrance, as shown, and tend to dissipate as the flow travels up through the tubes. The hottest tubes, illustrated with the whitest contours, are found on the symmetry plane at a location approximately 40% to 45% of the distance from the outer plenum wall to the center of the divider plate. Cold flow (return) tubes are indicated by the dark contours (black) on the figure. Most of the cold return flow tubes are at a uniform temperature equal to the secondary side temperature. At this level, 7 inches above the inlet plenum, some of the cold tubes have a small amount of hot flow present that blurs the boundary between the hot and cold tubes (see Fig. 12). This occurs on tubes closest to the hot tubes where individual tube flow rates are small. Hot gas rises into these tubes just enough to show up in these lower levels before it returns to the inlet plenum. The relatively large cross section of individual tubes resulting from the tube bundle modeling approach contributes to this phenomenon. A small recirculation cell appears to form at the bottom of some of the tubes. These slightly elevated temperatures in the cold tubes are not seen at higher elevations in the tube bundle. The boundary between the hot and cold tubes is determined from the data by looking at flow velocities, not temperatures, higher (7 m, 275 in.) in the tube bundle.

A major benefit of the CFD predictions is the ability to determine the range of temperatures within the tubes. System code predictions indicate only the mixed mean (a single hot tube) temperature. An indication of the range of temperatures within the hot tube region is clearly visible in the contour variations of Figure 12. To quantify this range, the temperatures are normalized over a range from 0 to 1 as described earlier. This range is broken down into 5% (0.05) intervals and the fraction of tubes within each interval is totaled. Figure 13 shows the results for cases h1 and h5 along with the 1/7th scale predictions. The predicted tube entrance flows are all within the range of normalized temperatures between 0.1 and 0.55. The 1/7th scale results have the same overall range as the full-scale predictions but show a slight increase in the average temperature. This result seems to indicate a slight decrease in the normalized temperature for the full-scale results compared to the 1/7th scale predictions. Case h1, which has more tubes in upflow and a lower average temperature, also shares the same overall range with case h5. Case h1 shows a slight decrease in the average normalized temperature value compared to case h5. The differences between all results are considered small. These data provide an enhanced view of the tube temperatures and provide much more information than a simple mass-averaged value. This type of result provides a means to study tube integrity for specific numbers of tubes at specific temperature ranges. The number of ranges used is considered adequate for this type of analysis.

5.3 Low-Temperature Scaleup Predictions

A set of lower temperature predictions is completed at a time in the transient when the hot leg hot temperature is close to 1000 K (1341 ^{o}F). The difference between the hot leg temperature and the secondary side temperature (total driving temperature difference) is 274 K. These lower temperature boundary conditions (I) result in reduced temperatures and temperature differences throughout the model. A comparison of these results with the high-temperature case provides a means to establish the sensitivity of the results to the temperature of the system.

Similar to the previous analysis, seven predictions (I1 through I7) are completed to quantify the effect of the heat transfer rate on the results. Case I1 corresponds to the lowest heat transfer rate and case I2 corresponds to the highest heat transfer rate. Similar trends observed for the high temperature cases are observed for these cases. Normalized temperature drops are shown in Figure 14 for cases I1 and I5 compared to the high temperature cases h1 and h5 and the two points in time from the SCDAP/RELAP5 predictions. Only I1 and I5 are shown to make the figure clearer. The heat transfer coefficients and porous media tube bundle assumptions for cases I1 through I7 are the same as in cases h1 through h7. The resulting normalized drop in the tube bundle temperatures for the hot flow tubes indicates relatively good agreement between the corresponding heat transfer rates (i.e., h1 vs. I1, and h5 vs. I5). The lower temperature results (I1, I5) are slightly above the corresponding high-temperature predictions, indicating a slightly reduced normalized rate of temperature drop for the lower temperature case. The SCDAP/RELAP5 predictions also indicate this trend. For the lower temperature point in the SCDAP/RELAP5 prediction, the results indicate a slightly higher normalized temperature and slightly lower rate of decline. As noted earlier, case h5 (and I5) correspond with the rate of normalized temperature drop from the 1/7th scale predictions completed in Reference 4. Cases I1 and h1 are most similar to the SCDAP/RELAP5 temperature drop predictions that are assumed to be prototypical.

Table 6 lists the low-temperature results for the scaleup model. Generally speaking, the trends are similar to the high-temperature case and a full discussion of the results is not repeated here. The percentage of hot tubes is minimum for the heat transfer rates associated with case I4 and I5. Case I5 corresponds to a heat transfer rate that reduces the normalized temperature in the hot flow tubes to the secondary side value at the top of the tube bundle. Increasing or decreasing the heat transfer rate from these levels results in an increased percentage of hot flow tubes. The temperature entering the hot flow tubes is relatively constant over the lower heat transfer rate range and then increases for cases I5, I6, and I7. This trend is also observed in the previous high-temperature runs. Tube bundle mass flow is highest for the lowest heat transfer rates and tails off at the higher values. The recirculation ratio shows the same trend. The recirculation ratio is at a level 0.25 lower than the results for the high temperature predictions for the lowest heat transfer rates. The recirculation ratios are essentially equal for the low- and high-temperature cases at the highest heat transfer rates. These differences are not considered to be significant in light of the overall modeling uncertainties.

The mixing fractions for the low-temperature cases are an average of 0.1 higher than the earlier predictions at the higher temperatures. The overall trend in this value with variations in the tube heat transfer rate is also similar to the high-temperature cases. The values are generally uniform near 0.95 with the exception of slightly elevated predictions of the mixing fraction for cases I5 and I6. The significance of these elevated mixing fractions in this range of heat transfer rates is unknown but does occur for both the high- and the low-temperature cases.

The mixing fraction predicted for cases I5 and I6 is larger than 1.0. This highlights a limitation in the simplified mixing model formulation from Reference 6[6]. A mixing fraction larger than 1.0 indicates that the bulk average tube entrance temperatures are lower than the mixed mean inlet plenum temperature as defined in the mixing model. The temperatures predicted for the tube entrance temperatures in these two cases are a few degrees below the predicted values for the mixed mean inlet plenum temperature.

19

5.4 Significance of Scaleup Predictions

The main purpose of the scaleup analysis is to isolate the effect of scale in the comparison of the $1/7^h$ scale predictions and full-scale predictions. In addition, these predictions help to quantify the significance of the secondary side heat transfer rate. These two goals are accomplished. The full-scale predictions are similar to those from the $1/7^h$ scale tests when the effective heat transfer rates from the tube bundle are similar. The full-scale simulations predict a slightly larger mixing fraction and associated lower normalized tube entrance temperatures. For this specific geometry and for similar normalized heat transfer rates, the $1/7^{th}$ scale tests provide a good representation of the full-scale behavior. The effect of scale on the fluid dynamic behaviors considered is minimal. This is not too surprising since some key scaling parameters from the $1/7^{th}$ scale tests are within an order of magnitude of (or closer to) the full-scale parameters (Ref. 4).

The second result from these predictions confirms that tube bundle heat transfer rate is a key parameter affecting the flow and mixing behavior. Although the mixing fraction itself is not significantly affected by variations in this parameter, the recirculation ratio and the resulting tube entrance temperatures are strong functions of the tube bundle heat transfer rate. This result highlights the need to accurately model the secondary side heat transfer in this type of analysis. The lowest heat transfer rate used in these predictions, in cases h1 and l1, is most representative of the specific SCDAP/RELAP5 predictions for the ZION TMLB' station blackout transient used for this analysis.

Finally, the predictions provide quantitative results for the temperature variations across the hot flow tube section. The predicted range of temperatures provides a better indication of the true challenge to the SG tubes. The magnitude and extent of the hottest temperature range is now available.

6 PROTOTYPICAL WESTINGHOUSE STEAM GENERATOR ANALYSIS

The following predictions were done to study the flow and mixing phenomena in a prototypical Westinghouse-designed steam generator inlet plenum and to determine the significance of the geometrical difference between the prototypical geometry and the 1/7[h] scale facility. A geometry based upon the primary side of a model 44 steam generator was used. As indicated in Figure 2 and Table 1, this geometry is not completely the same as the scaleup model studied earlier. The nonsymmetric hot leg orientation is the most significant difference. The dimensions in the inlet plenum region are also different.

Figure 15 shows a cross section of the inlet plenum on a vertical plane parallel to the hot leg for the scaleup model and the prototypical Westinghouse geometry. A significant difference is the distance from the top of the hot leg to the tube sheet entrance. This distance, which can be thought of as a mixing length for the rising hot plume, is shorter for the prototypical model 44 design. The hot gas exiting the hot leg does not have to travel as far to reach the tube sheet in the model 44 design as in the scaleup model. The distances from the top of the hot leg to the tube sheet entrance highlighted on Figure 15 are approximately 1.06 m (42 in.) for the scaleup model and 0.81 m (32 in.) for the prototypic design.

The boundary conditions from Table 3 also apply to this analysis. A single high- and low-temperature case are completed with one value for the secondary side heat transfer rate. Boundary conditions, mesh size, and modeling options are selected to be the same as the scaleup model. The only significant difference between the two models is the geometry.

Results from these analyses are fundamentally different from the scaleup predictions completed earlier. The major oscillations predicted for the rising hot plume in the inlet plenum are the key difference. The flow pattern changes substantially during a typical oscillation and the resulting mixing parameters also oscillate. A simple steady-state solution was not obtained. A transient solution technique is utilized with a time step small enough to ensure time accuracy for a period of time long enough to cover multiple oscillations. The solutions were monitored to ensure a repeatable pattern emerged. It required an order of magnitude more computer time and analyst time to complete this process than in the scaleup analysis completed earlier.

Many solution parameters oscillated and the periods for these oscillations are not necessarily the same for different parameters. A mass imbalance for the entire model is one parameter used to monitor the solution convergence and this parameter is used to define an average oscillation. The mass imbalance oscillated with a period of approximately 25 seconds. Inlet plenum plume oscillations are complex but appear to have a period equal to the mass imbalance. However, due to the limitations and assumptions of the tube bundle modeling approach, this duration is not necessarily indicative of the prototypical behavior. The tube bundle is expected to affect the plume oscillations.

To ensure the convergence of the solutions, several key parameters are monitored during the transient solution. Full data sets are saved at 1-second intervals and these data are used to determine the average and standard deviation of parameters during a typical oscillation.

6.1 Prototypical Westinghouse High-Temperature Predictions

Figure 16 shows contours of temperature on a vertical plane parallel to the hot leg with the region of the hot plume shown at 3-second intervals during a typical oscillation. The plume is continually moving. During some intervals, such as the interval from 15 to 24 seconds, the plume generally rises to the tube sheet more or less the same as in the scaleup model predictions. This pattern is significantly different than during the interval from 0 to 12 seconds when the plume appears to be pushed back from the tube sheet. At around 12 seconds the plume is nearly stagnated momentarily. This behavior is also evident in the temperature contours on a plane near the tube entrance.

Figure 17 provides contours of temperature on a horizontal plane 7 inches above the tube sheet entrance. Temperature contours over the region surrounding the hottest tubes are given at 3-second intervals to illustrate the changes in the magnitude and location of the hottest tubes. The results in Figure 17 correspond directly to the results in Figure 16. The lateral variations in the location of the hottest tubes are evident in the figure. At 12 seconds, when the plume appears to be momentarily stagnated in Figure 16, the tube entrance temperatures are reduced as illustrated in Figure 17. This behavior is fundamentally different than predicted for the scaleup model. The number of tubes that see the hottest temperatures is increased in this model. However, for a given set of tubes, the temperature varies significantly from the hottest value to more moderate values during a typical cycle.

At this point, the effectiveness of the tube bundle heat transfer rate is examined to facilitate a comparison with the scaleup predictions. It is desirable to compare the results only where the tube bundle heat transfer rates are consistent since this parameter can have a significant effect on the results. This comparison is made by looking at the temperatures in the tubes and the rate at which they decrease along the flow path. Figure 18 shows the normalized temperature for the flow in the tube bundle. Predictions h1 through h7 from the scaleup predictions are shown as the solid circles connected by lines and the SCDAP/RELAP5 results are indicated for a reference. The prediction for a prototypical Westinghouse design is shown as the larger hexagons. The temperature profile starts at a higher initial temperature than the scaleup predictions, which is an indication of less inlet plenum mixing. The shape of the temperature profile appears most similar to case h4 although the model 44 predictions are at a consistently higher temperature level. Near the upper portion of the tubes, the profile is closest in magnitude to case h3. This prediction for the model 44 steam generator is compared with the scaleup cases h3 or h4. Note that the results for cases h3 and h4 are similar.

The tube entrance temperatures are collected into ranges and a histogram is created for the model 44 steam generator results. For each temperature range in the histogram, an average value and a standard deviation are determined. A total of 25 data sets, spaced at 1-second intervals, are used to determine these values. The results are compared to case h4 from the scaleup model in Figure 19. The highest normalized temperature range is 0.65 to 0.7 for these predictions. This is somewhat higher than the scaleup prediction. Only a few tubes see this temperature and the standard deviation range demonstrates that the tubes do not stay at this level. As noted earlier, the model 44 design was expected to result in less mixing than the scaleup model due to the reduced plume travel in the inlet plenum (see Fig. 15). Less mixing for the model 44 design is also evident on Figure 18, where the model 44 results start at a higher normalized temperature than in all of the scaleup predictions.

22

The histogram alone does not provide a fair comparison between the model 44 and the scaleup predictions of the tube entrance temperatures due to the nature of the oscillations predicted for the model 44 design. The hottest region of the tube bundle varies with time and space as noted earlier and illustrated by Figure 17. The scaleup predictions indicated a consistent peak tube temperature value and location. The eight hottest tube regions from Figure 17 are isolated and labeled a through h as shown in Figure 20. Next, the mass averaged temperature entering each of these sections is plotted as a function of time. The highest tube temperature on the plot occurs in section a at 0 seconds. This same tube region has the lowest temperature within this group of tubes 12 seconds later. That the hottest tube has a normalized temperature of 0.65 as indicated by the histogram (Fig. 19) is a conservative assumption. It may be more appropriate to apply a time-varying temperature to the tubes to account for the movement of the plume. The current tube bundle model, however, is not designed to have the same time response as a prototypical tube bundle and the actual period of the oscillatory behavior is unknown.

Results for the prototypical model 44 analyses are listed in Table 7 along with a set of predictions from the scaleup model. The selected parameters are obtained at 1-second intervals. An average value and a standard deviation are determined from a typical oscillation consisting of 25 data points. The scaleup predictions listed in Table 7 are selected because they have a similar rate of temperature decrease in the tubes (see Fig. 18).

The prototypical model 44 predictions indicated that 44.4% of the tube bundle carried hot flow. This is slightly higher than the scaleup predictions. The oscillating nature of the rising hot plume could help to spread the flow to a larger number of tubes. One feature of this prediction is the stability of this particular value. Although the solution oscillates and most parameters have an associated standard deviation, the percentage of tubes carrying hot flow is nearly constant. It appears that the tube flows have a certain level of inertia that keeps them consistent even though the flow conditions at the tube sheet are oscillating. Note that the average tube residence time, or the time it takes for the flow to pass from the inlet plenum to the outlet plenum and back, is approximately 240 seconds. This is roughly 10 times the period of the inlet plenum oscillations. The relatively short oscillations apparently cannot affect the bulk tube flow rate. Another notable point is that the tube residence time of 240 seconds refers to the CFD model with the porous tube bundle model. This model attempts to preserve the tube bundle mass flows but does not predict the actual flow velocities. For the given mass flow rate, the average tube bundle residence time for the prototype would be approximately 80 seconds.

The reported values for the hot leg mass flow and temperature are close to the applied boundary conditions and are essentially equal to the scaleup model conditions. The temperature of the return flow in the hot leg is affected by the mixing in the inlet plenum and other predicted parameters. This value, 924 K (1203 °F), matches the scaleup prediction for h4.

A key prediction is the temperature of the flow entering the tube bundle. These values were considered in the discussion of Figures 19 and 20. The table lists the mass-averaged value for the entire tube bundle flow. This value, 1049 K, is 16 K higher than the scaleup result for case h4. This is consistent with the prediction of less mixing for the model 44 design, as noted earlier. The mixing fraction is 0.8 +/- .06, which is less than the 0.87 value predicted for the scaleup model.

Another key prediction is the tube bundle mass flow. This value is consistent between the prototypical model 44 prediction and the scaleup prediction. Likewise, the recirculation ratio is consistent for these two models. It appears that the predictions for the prototypical model 44 design are most consistent with the scaleup prediction for case h4. The areas of significant variation are the lower mixing fraction (with the associated higher tube entrance temperature) and the slightly larger number of hot flow tubes.

6.2 Prototypical Westinghouse Low-Temperature Predictions

The low-temperature predictions generally resulted in the same behavior as the high-temperature predictions described above. Similar solution oscillations are predicted with the mass imbalance oscillating with a period of approximately 26 seconds. To facilitate a comparison of the results with the scaleup predictions, the effectiveness of the tube bundle heat transfer rate is gauged by looking at the temperatures in the tubes and the rate at which they decrease along the flow path. Figure 21 shows the mass-averaged normalized temperature for the hot flow in the tube bundle as a function of the normalized height from the tube sheet entrance to the top of the tube bundle. Predictions I1 through I7 from the low-temperature scaleup predictions are shown as the solid circles connected by lines and the SCDAP/RELAP5 results are indicated for a reference. The low-temperature predictions for a prototypical Westinghouse design are shown as the larger hexagons. The temperature profile starts at a higher initial temperature than in the scaleup predictions, which is an indication of less inlet plenum mixing. The shape of the temperature profile appears most similar to case I4 although the model 44 predictions are at a consistently higher temperature level. Near the upper portion of the tubes, the profile is closest in magnitude to case I3. The normalized temperatures for this low-temperature prediction are slightly higher than in the high-temperature predictions for the model 44 design. This trend of higher normalized temperatures at lower hot leg temperatures (ie, earlier in the system heatup) is consistent with the SCDAP/RELAP5 predictions and the scaleup results discussed earlier.

Specific results from the model 44 steam generator prediction are compared with the scaleup predictions from cases I3 and I4 in Table 8. Compared to the scaleup predictions, the model 44 results differ mainly in the lower mixing fraction and the resulting higher tube entrance temperatures. The mixing fraction of 0.81 is almost identical to the value for the high-temperature case (0.80). This comparison with the scaleup model at low temperatures gives the same results as the comparison at high temperatures. Tube bundle flows and recirculation ratios are similar but the reduced mixing results in higher tube inlet temperatures.

A histogram of the tube entrance temperatures for the high and low temperature cases is given in Figure 22. The low-temperature case extends over the same general range as the high-temperature prediction. Both peak near a normalized temperature range between 0.65 and 0.7. The low-temperature scaleup results are not shown in the figure but are similar to those in Figure 19.

To quantify the variations in the tube entrance temperatures, a history of the mass-averaged temperature entering eight of the hottest tube sections is given in Figure 23. This is similar to Figure 20 for the high-temperature predictions. Normalized temperatures oscillate between 0.4 and 0.67. These data provide an indication of the variation in the temperatures expected to enter the tube bundle in the hottest tube region.

24

6.3 Significance of Predictions for the Prototypical Westinghouse Design

The predictions for a prototypical Westinghouse design provide insights into the effect of the geometrical differences between a Westinghouse steam generator and the Westinghouse 1/7th scale test facility.   Table 1 and Figure 2 along with Figure 15 highlight the differences between the geometry of the facility and the prototype. The effect the geometry variation is quantified by comparing predictions for the scaleup model and the prototype design with similar boundary conditions applied.

The significant differences predicted are the reduced mixing in the inlet plenum and the oscillatory plume behavior predicted for the nonsymmetric model 44 geometry. Reduced mixing is attributed to the shorter distance between the hot leg and the tube sheet. This reduced mixing shows up in mixing coefficients near 0.8 and in the resulting higher tube entrance temperatures (compared to the scaleup model).

The predicted hottest tube section varies in magnitude and location as the plume oscillates during the transient solution. Scaleup predictions indicated steady plume and solution behavior. The plume oscillations in the model 44 predictions provide the opportunity for a more detailed analysis of the hottest tubes. A single hottest tube region is not predicted. Instead, a larger number of tubes will see the hottest temperature. The temperature at a given location, however, will not remain at the highest level. A significant variation in the temperature at a given location is possible. Figures 20 and 23 provide insights into this behavior.

Despite the differences, the solution does have some major similarities. The percentage of tubes carrying hot flow is predicted to be close to the scaleup model predictions. In addition, the predicted tube bundle mass flows and the recirculation ratios are consistent. The change in the mixing fraction is considered significant but the difference is not considered enormous. Significant mixing of the inlet plenum hot plume still occurs. The overall effect of the decreased mixing fraction and the resulting increase in the tube entrance temperatures is mitigated somewhat by the oscillations of the plume. A detailed tube integrity analysis will be needed to quantify the significance of the variation in the mixing parameters. This type of analysis is beyond the scope of this report.

7 ANALYSIS WITH LEAKAGE FROM TUBES

Tube leakage from the bundle is a concern if the leaking tube or tubes pulls unmixed flow through the leak. The concern is whether the unmixed hot leg gas will heat the leaking tube beyond expectations and lead to an enlargement of the leak or a complete tube failure. The issue of leaking tubes was not addressed in the 1/7th scale experiments.

The effect of the leak on the tube entrance temperatures is of primary concern. Four specific leak locations are defined as shown in Figure 24. Three individual leaks are positioned on the symmetry plane and labeled leak 1, leak 2, and leak 3. Leak 2 is positioned near the hottest region of the tube bundle. This leak is most likely to draw hot gas from the rising plume. Leaks 1 and 3 are positioned on either side of this leak near the edges of the hot tube region (see Fig. 24). The fourth leak considered is a distributed leak spread out over the entire tube sheet in a pattern that affects 21 of the 216 tube regions. All leaks are established on the side of the leaking tube region that faces the hot leg. The leak begins at the top of the tube sheet (0.8 m [31.5 in.] above the inlet plenum) and extends upward 0.2 m (7.9 in.). Figure 24 indicates the leak locations as black lines on the edge of the leaking tube region. The arrows in the leaking tube region point toward the leak. The boundaries of the adjacent tube regions surrounding the leak are not affected by the leak.

Tube leakage predictions are based upon the scaleup model described in Section 5. The high-temperature boundary conditions from case h5 are applied to the model in addition to the set of tube leakage boundary conditions. The only difference between this model and the model used for case h5 is the introduction of the leaking boundary condition. Table 3 lists specific conditions for the high-temperature case. Direct modeling of the choked flow for specific leak geometries would require a more detailed tube bundle model and is not attempted. Leak rates are established by specifying the mass flow rate from the given leak. Leak geometry is not critical to this approach. To estimate the mass flow rate from leaks, a typical leak rate model is used to define a hole size associated with a leak rate of 150 gallons per day (gpd) standard operational conditions. This hole size is used to estimate a mass flow rate for superheated steam at the severe accident conditions of interest. The desired mass flow is then established across the specific leak boundary in the model.

Assuming leak conditions at 1150 K (1610 °F) and 2400 psi (severe accident conditions), the hole size determined above for the 150 gpd leak results in a superheated steam flow rate of 0.0014 kg/s. The 0.0014 kg/s leakage flow is of the same order of magnitude as the overall mass imbalance in the CFD model and is insignificant in light of the nearly 10 kg/s tube bundle flow rates. Larger leak rates are selected to provide a meaningful leak rate analysis. Values of 0.014, 0.14, 1.4, and in one case 2.8 kg/s are selected for these predictions to provide a wide range of conditions. These conditions span the range of leak hole sizes consistent with standard operating condition leak rates ranging approximately from 1500 gpd to 200 gallons per minute (gpm). For the purposes of this report, the leaks are discussed in terms of the mass flow rates exiting the leak at severe accident conditions. Relating these mass flow rates directly to specific leak geometries is beyond the scope of this report.

7.1 Predictions for Leak 1

Leak 1 is positioned closest to the hot leg as shown in Figure 24. This leak is just above the hot leg nozzle near the edge of the hot tube region. A quick look at the results shows that the hot plume is not pulled into the leaking tube. This is a significant finding. The hot plume actually moves further from the leaking tube position as the leak rate increases. Figure 25 shows temperature contours on the symmetry plane for leak 1. In the region just below the tube sheet, the contours are provided for each of the four leak rates to give an indication of the movement of the hot plume. The hottest region of the hot plume and its axis are indicated by the white arrows. These white arrows are indicative of the plume path. The results for the 0.014 and 0.14 kg/s leak rates are essentially the same and do not differ significantly from case h5 with 0 leakage. The plume moves out slightly further from the hot leg when the leak rate is increased to 1.4 kg/s and this trend continues as the tube leakage rate is increased to 2.8 kg/s. The leak location is indicated by the white line just above the top of the tube sheet on the second tube region from the left (see Fig. 24). The movement of the hot plume away from this particular leak as the leak rate is increased is due to the reduced return flow from the steam generator tubes. Flow passing from the return flow tubes and into the hot leg tends to resist the forward progress of the hot plume from the hot leg. The reduction in this return flow allows the hot plume to enter the inlet plenum with less resistance and therefore extend further into the inlet plenum. The return flow is reduced as the leak rate is increased since the hot leg flow rate is fixed by the boundary condition.

Figure 26 further illustrates the changes in the plume location by looking at temperature contours 7 inches above the tube sheet entrance for each of the four leak magnitudes. The leaking tube is indicated by the white line (second tube region from the right on the symmetry line in these figures). The hottest region of the flow is outlined with a black oval. The results for the 0.014 and 0.14 kg/s leak rates are essentially the same. The center of the hot test region is 3.5 tube region widths from the leak location. All tube regions have the same width and the tube region boundaries are visible on Figure 26. For the 1.4 kg/s leak rate, the center of the hottest region moves slightly away from the leak (approximately 3.8 tube widths from the leak). Finally, the hottest tube region for the 2.8 kg/s tube leak case is located 4.5 tube region widths from the leak. This movement of the hottest region is consistent with the movement of the hot plume observed in the symmetry plane temperature contours on Figure 25.

Clearly the hot plume is not pulled into the leaking tube in this example. At leak rates above 1.4 kg/s, the leaking tube does reduce the return flow from the cold tubes and does begin to impact the inlet plenum flows and mixing. Predicted normalized temperatures entering the tube sheet are plotted in the form of a histogram on Figure 27 for each of the four leak rates. Leak rates of 0.014 and 0.14 kg/s are essentially identical. A slight increase in the number of tubes in the 0.5 to 0.55 normalized temperature range is predicted for the 1.4 kg/s leak rate. For a leak rate of 2.8 kg/s, the hottest tube normalized temperatures increase into the 0.6 to 0.65 range. It is noted that the hot leg temperature used to normalize these results is lower for the 2.8 kg/s case. The absolute temperature of the hottest tube region, however, is still higher for the 2.8 kg/s case. In summary, as the leak rate is increased to 2.8 kg/s, the temperature of the hottest tube region increases and this region moves further away from the hot leg.

Table 9 summarizes some of the key mixing parameters for these predictions along with some leak-specific data. Case h5 from the scaleup analysis is included to provide the baseline result with zero tube leakage. Although some of the trends are significant, it is important to understand the assumptions and limitations of the model before drawing conclusions (see Section 7.5 below). The 0.014 and 0.14 kg/s tube leakage rate predictions are essentially the same as for case h5 from the scaleup analysis. The following discussion is focused on the 1.4 and 2.8 kg/s leak rates.

The number of tubes carrying hot gas is not affected by the amount of the leak. The differences seen in the table represent a 1 tube region variation in the symmetry model (2 out of 216 total tube regions) and no trend is observed. The hot leg flows do show some variation at the higher leakage rates. The trends are affected by the assumption of a fixed hot leg inlet mass flow for all leak rates. For instance, the hot leg hot flow temperature (observed midway between the steam generator and the vessel) is reduced by approximately 30 K when the leak rate is 2.8 kg/s. This is due to significant entrainment of colder flow in the hot leg. This entrainment shows up in the increased hot leg mass flow (predicted at the center of the pipe) for this case. In a reactor system, the hot leg mass flow is also expected to increase when the tube leakage increases since the leaking leg would pull some additional hot flow from the vessel. A temperature reduction in the hot leg flow is not expected in this case. It is important to understand the impact of the fixed boundary conditions used in the CFD model and how they might change the behavior of the system.

A surprising result is the slight reduction in the mass-averaged temperature of the flow entering the tube sheet for the 2.8 kg/s leak rate case. This reduction is small but is not expected based upon the increased temperature range noted on Figure 27. Although some hotter temperatures enter the tube sheet as shown on Figure 27, the leak itself is pulling in relatively cooler flow at leak 1. This has the tendency to reduce the mass-averaged temperature entering the tube sheet. In addition, the mass-averaged temperature of the hot leg flow is reduced at the highest tube leakage rate as described above. Overall, these variations in the temperature entering the tube sheet are not too significant.

The mass flow entering the tube sheet increases almost as much as the tube leakage rate. A portion of the flow exiting the leaking tube comes from the outlet plenum for the two highest tube leakage rates. When the tube bundle hot tube flow rate is computed above the leak location, there is no significant variation with increased leakage rate. This nearly constant tube bundle mass flow rate is used to determine the recirculation ratio. A decrease in the recirculation ratio at the highest tube leakage rate is due mainly to the increase in the predicted hot leg mass flow used to determine the recirculation ratio.

The mixing fraction is determined from the hot leg hot temperature, hot tube entrance temperature, cold tube return flow temperature, and the recirculation ratio. A slight increase in the mixing fraction is predicted for the highest tube leakage rate case. This is the result of a decrease in the recirculation ratio used for the determination of the mixing fraction. However, the method used to determine these values must be considered. The assumptions used to determine the mixing fraction do not consider tube leakage. A revised mixing model is beyond the scope of this report.

Finally, Table 9 lists some specific data from the tube leak. The mass flow exiting the boundary is shown along with the mass flow entering the leaking tube region from the inlet plenum. For the two lowest leakage rates, the mass flow entering the leaking tube is more than the leak rate and part of the flow continues on to the outlet plenum. For the two highest tube leakage rates, the mass flow entering the leaking tube is less than the tube leakage rate. A portion of the leak flow comes from the outlet plenum. The leak flow temperature is the mass-averaged temperature of the flow exiting the system at the leak boundary. This value decreases at the higher leakage rates due to the portion of relatively cold gas that comes from the outlet plenum.

7.2 Predictions for Leak 2

Leak 2 is positioned near the center of the hot plume region (see Fig. 24). This leak position is most likely to draw the hot plume into the leaking tube region. Leak rates of 0.014, 0.14, and 1.4 kg/s are applied to leak 2. Figure 28 illustrates temperature contours on the symmetry plane with the region around the tube sheet entrance pulled out for each of the three leakage rates. The results for 0.014 and 0.14 kg/s are very similar to each other and the zero leakage case. With a leakage rate of 0.14 kg/s, the hot plume moves slightly further from the hot leg but is not pulled up into the leaking tube region. The leaking tube region pulls gas from the inlet plenum but there does not seem to be a preference for pulling in the hottest gas.

The normalized tube entrance temperatures are provided on Figure 29 for the three leak rates at leak 3. The results are essentially the same as those for leak 1 (see Fig. 27). The temperature increase at a leak rate of 1.4 kg/s is very small. The steam reaching the leak still passes through the inlet plenum with a similar amount of mixing to the zero leakage case.

Tabulated results for leak 2 are provided in Table 10. Generally speaking, the results are very similar to those described for leak 1 above. A complete discussion will not be repeated here. One difference for leaks at position 2 is the temperature of the steam that passes through the leak. Leak 2 draws gas from near the hottest region of the plume. As expected, the tube leakage temperature reported in Table 10 is higher than the temperature predicted for leak location 1. Leak 2 slightly reduces the amount of hot gas available to drive the circulation in the tubes. A slight decrease in the tube heat loss is predicted for the 1.4 kg/s leak rate. In addition, there is a reduction in the net tube bundle mass flow (above the leak location).

With the exception of the temperature of the steam pulled into the leak and the effects discussed above, the leak 2 predictions are very similar to the leak 1 predictions. Mixing is consistent with the zero leakage case for leak rates up to 1.4 kg/s. The recirculation ratio and the mixing fraction are not significantly affected.

7.3 Predictions for Leak 3

Leak 3 is near the divider plate on the symmetry plane as shown on Figure 24. Leak rates of 0.014, 0.14, and 1.4 kg/s are applied at this location. The predictions for the plume location and the tabulated mixing parameters are essentially the same as those for leak 1. A complete description of these results is not repeated here. The conclusion of this analysis is that the position of the leak, outside of the hot plume region, does not significantly impact the results. The hot plume is not drawn to the leak and the only impact on mixing is from the reduction in the return flow to the vessel.

7.4 Predictions for a Distributed Leak

The distributed leak consisted of 21 tubes arrayed in a pattern across the entire tube sheet. Almost 10% of the tubes are leaking. Leakage rates of 0.14 and 1.4 kg/s are distributed equally across the 21 leaking tubes. The detailed results are not provided here because of the complete similarity to the predictions for leak 1. The histogram of normalized temperatures showed no difference between the 1.4 kg/s leak rate and the zero leak case (h5). The mixing parameters and other predictions are essentially the same as those reported in Table 9 for leak 1.

7.5 Limitations of Leakage Predictions

This leakage analysis does not account for any changes in the overall system behavior that are expected in the event of a significant tube leakage. For instance, it is expected that the hot leg mass flow leading to the leaking generator would increase as the tube leakage increases. No system feedback is accounted for in these predictions. All boundary conditions are held constant from case to case and are consistent with case h5. The larger leak rates could have a more significant impact on the tube entrance temperatures if the hot leg mass flow is increased as a result of the leak. The general limitations outlined in Section 4 of this report are also relevant to these predictions.

7.6 Summary of Tube Leakage Predictions

Tube leakage from a variety of locations has been considered and a few general conclusions can be reached. Tube leakage up to 1.4 kg/s does not significantly impact the inlet plenum mixing. The hot plume is not predicted to be diverted into the leak and the leak does not cause a bypass (with no mixing) of the inlet plenum. At a leak rate of 2.8 kg/s, the hot plume enters the inlet plenum with less resistance and intersects the tube sheet further from the hot leg nozzle. The reduced resistance and slightly reduced mixing predicted for the leakage rate of 2.8 kg/s is the result of the reduced return flow to the vessel. A portion of the flow that normally returns to the vessel during this type of flow pattern exits the system through the leaking tube.

8 COMBUSTION ENGINEERING PLANT STEAM GENERATOR ANALYSIS

The Combustion Engineering (CE) steam generators are not geometrically similar to the Westinghouse designs and there is a concern whether the $1/7^{th}$ scale Westinghouse test data is applicable to CE plants. The effect of the geometry on the inlet plenum mixing is studied by considering a typical steam generator from a CE nuclear power plant. A geometry based upon the primary side of a replacement steam generator (circa 2001) for the Calvert Cliffs nuclear power plant is used for these analyses. As indicated in Figure 2 and Table 1, this geometry is larger and of different proportions than either the scaleup model or the model 44 design studied earlier. The relative size of the hot leg in proportion to the inlet plenum and the distance from the hot leg to the tube sheet are the most significant differences in this design.

Figure 30 shows a cross section of the inlet plenum on a vertical plane parallel to the hot leg for this particular CE plant steam generator and a prototypical Westinghouse geometry (model 51 steam generator). The CE design is not as high and the hot leg is closer to the tube sheet. The distance from the top of the hot leg to the tube sheet entrance can be thought of as a mixing length for the rising hot plume. This distance for the CE design, 0.351 m (13.8 in.) is less than half of the distance noted for the model 44 design on Figure15. The significance of this difference is magnified by the fact that the hot leg diameter in the CE design (42 in.) is significantly larger than in the Westinghouse design (29 in.). Assuming the hot plume entering the inlet plenum has a diameter that is 25% of the hot leg diameter, the mixing length for the CE design is roughly 1.3 plume diameters. This is compared to a mixing length of approximately 4.4 plume diameters for the Westinghouse design. Although the flow topology in these two designs is different and a simple mixing length is not sufficient to describe the phenomena, these numbers do provide a good indication of the expected differences between the Westinghouse and CE designs.

The boundary conditions for this analysis are listed in Table 11. These conditions come from a SCDAP/RELAP5 analysis of the Calvert Cliffs nuclear power plant during a station blackout transient. A single high- and low-temperature case are completed with one value for the secondary side heat transfer rate. Where possible, modeling options and the general mesh sizes are selected to be consistent with the Westinghouse analysis completed earlier.

Results from these analyses are significantly different from the Westinghouse predictions. The inlet plenum mixing is reduced due to the reduced height of the inlet plenum. Figure 31 shows the temperatures on a vertical symmetry plane of the model for the high-temperature case. The region of the hot plume in the inlet plenum is enlarged to provide details of the temperature contours in this region. The hottest temperature region indicated by the white contour region extends from the vessel exit to the tube sheet entrance plane. Although mixing reduces the size of the hottest region, it appears that a small portion of the hot leg flow reaches the tube sheet entrance without significant temperature reduction. This relative lack of mixing is consistent for both the high- and low-temperature boundary conditions.

The steady-state solutions reached consistent values without significant oscillations. This solution behavior is similar to the scaleup model predictions completed earlier. It is unclear whether the symmetric design of the steam generator inlet plenum is responsible for the steady solution behavior or whether the assumption of symmetry in the numerical model is limiting the oscillations. The solutions are obtained with a transient solver using fixed boundary conditions.

33

To ensure convergence of the solutions, several key parameters are monitored until a steady value is reached. An overall mass and energy balance is also monitored for convergence.

8.1 Combustion Engineering Plant Predictions

Figure 31 makes it clear that the mixing for the CE design is lower than the mixing predicted in the Westinghouse analysis completed earlier. Looking at the temperatures entering the tube sheet reinforces this conclusion. Figure 32 is a histogram of the normalized temperatures entering the tube sheet. The predictions for the high- and low-temperature boundary conditions are included. These results clearly show the impact of the reduced inlet plenum mixing when compared to Figure 19 for the Westinghouse predictions. The hottest tubes have a normalized tube entrance temperature near 0.95. This is only a 5% reduction from the average hot leg temperature. Although the tube sheet will reduce this temperature before the flow reaches the tubes, these temperatures are significantly higher than the predictions for tube entrance temperatures in the Westinghouse design. The results from the histogram are consistent with the temperatures observed in Figure 31 for the high-temperature case.

It is noted for completeness that the normalized temperature uses the mass-averaged hot leg temperature as a hot reference temperature. A portion of the hot leg flow is at temperatures higher than the mass-averaged value. A histogram of normalized temperatures in the hot leg would result in values greater than 1.0. This helps explain why the normalized temperatures are so close to 1.0 in Figure 32.

The mixing parameters for the CE plant predictions are provided in Table 12 for both sets of boundary conditions outlined in Table 11. The results show some variation with the different boundary conditions. The reduced mixing compared to the Westinghouse design is consistent, however, for both the high- and low-temperature case. Mixing fractions of 0.58 and 0.64 are predicted for the high- and low-temperature cases. The percentage of tubes carrying hot flow in the tube bundle ranges from 37% to 46%. This is a range similar to the Westinghouse predictions. The hot leg mass flows are significantly larger than the Westinghouse predictions due to the nature of the steam generator. The CE plant studied uses two large steam generators and the Westinghouse plant studied utilizes four loops with relatively smaller steam generators.

The predicted recirculation ratios are significantly lower for the CE plant predictions. The values range from 1.44 to 1.64. These lower values indicate that proportionately less flow enters the steam generators than in the Westinghouse predictions. The flow that enters the steam generators is the hot flow rising through the inlet plenum. A shorter path length through the inlet plenum entrains less fluid that can enter the tubes. The CE design has a relatively short path through the inlet plenum as noted earlier. Less fluid is expected to be entrained into the rising hot plume. This explains the relatively reduced tube-bundle to hot-leg mass flow ratio and the increased temperature of the flow entering the tubes in comparison to the Westinghouse design.

8.2 Significance of Predictions for the Combustion Engineering Design

The predictions for a prototypical CE design provide insights into the effect of the geometrical differences between a Westinghouse steam generator and a specific Combustion Engineering design. Table 1 and Figure 2 along with Figure 30 highlight the differences in the geometry of the CE design compared to the Westinghouse design. The biggest geometrical difference is the reduced distance from the hot leg nozzle to the tube sheet entrance.

Significantly less mixing is predicted for the CE design than in the Westinghouse predictions. The highest temperatures entering the tube sheet are very close to the hot leg temperatures. A small portion of the hot leg flow enters the tube sheet with only a small reduction in temperature. The impact of this reduced mixing on the tube integrity for a CE plant will have to be reevaluated in light of these predictions.

9 SUMMARY AND RECOMMENDATIONS

Building upon the successful comparisons with test data at 1/7th scale (Ref. 4), these predictions provide a unique look at steam generator inlet plenum mixing and entrainment during a severe accident scenario under a variety of full-scale conditions. The analyses provide a detailed look at the effects of geometrical variations, boundary conditions, and tube leakage on the mixing parameters of interest. The results provide an extension of the related 1/7th scale data into areas of interest to the USNRC and provide a basis for updating the parameters used in SCDAP/RELAP5 to account for the inlet plenum mixing.

The effect of scale on the inlet plenum mixing is analyzed first. This set of predictions are completed using a full-scale geometry with complete similarity to the 1/7th scale facility. Predictions at 1/7th scale and full-scale conditions are compared and indicate no significant effect of scale on the mixing. The recirculation ratio and the percentage of hot tubes are predicted to be the same as in the 1/7th scale test. The mixing fraction at full-scale conditions is slightly higher than the 1/7th scale value. Three conclusions are drawn from the scaleup analyses. The results are consistent over a significant variation in hot leg temperatures. The tube bundle heat transfer rate has a significant impact on the mixing parameters of interest. And finally, the 1/7th scale tests are indicative of the full-scale behavior when consistent secondary side heat transfer rates are applied to similar geometry.

The second issue addressed is the impact of the geometric differences between the 1/7th scale test facility and a prototypical Westinghouse model 44 steam generator. Predictions for a model 44 steam generator are completed using the same flow and heat transfer boundary conditions as the scaleup analyses described above. The model 44 predictions indicated significant variations in the plume intensity and location. On average, the mixing is less than the scaleup model indicates. The oscillations are thought to be a result of the nonsymmetric inlet plenum design. Oscillations in the scaleup model, however, could be diminished by the assumption of symmetry. The percentage of tubes carrying hot flow is relatively close for the two designs and the mass flows and recirculation ratios are consistent. Temperature variations in both time and space are available for a detailed tube integrity analysis.

The issue of tube leakage is addressed by considering leaks up to 2.8 kg/s and a variety of leak locations. Tube leakage up to 1.4 kg/s does not significantly impact the inlet plenum mixing. The hot plume is not predicted to divert into the leak and the leak does not cause a bypass (with no mixing) of the inlet plenum. At a leak rate of 2.8 kg/s, the hot plume enters the inlet plenum with less resistance and intersects the tube sheet further from the hot leg nozzle. The reduced resistance and slightly reduced mixing predicted for the leakage rate of 2.8 kg/s is the result of the reduced return flow to the vessel. Leakage rates of 0.014 and 0.14 kg/s were considered and had no significant impact on the results.

The final analysis looks at a sample Combustion Engineering (CE) plant steam generator. The geometry is noted to be considerably different than in the Westinghouse design with a hot leg positioned relatively close to the tube sheet entrance. Significantly less mixing is predicted for the CE design compared to the Westinghouse predictions. A small portion of the hot leg flow enters the tube sheet at a temperature close to the hot leg flow temperatures. The impact of this reduced mixing on the tube integrity for a CE plant will be reevaluated using SCDAP/RELAP5 analysis in light of these predictions.

The results from these analyses provide insights into the system behavior that can be used to estimate mixing parameters for SCDAP/RELAP5 analysis of the natural circulation flow scenario outlined in Figure 1. Some interpretation of these results is necessary due to the limited number of predictions and the assumptions and limitations of the modeling approach. In particular, the influence of tube bundle heat transfer rates on certain mixing parameters must be considered.

The key mixing parameters determined from experiments and code predictions that are needed to setup a SCDAP/RELAP5 analysis of the natural circulation flows are the tube flow split fraction, inlet plenum mixing fraction, recirculation ratio, and the percentage of core power deposited into the steam generators. This analysis does not independently predict the core power transferred to the steam generators since the boundary conditions at the hot leg entrance are fixed. Only the first 3 parameters mentioned will be addressed. In addition to the mixing parameters, specific details related to the location and temperature ranges of the hottest tubes are predicted. These data can be used to augment the one-dimensional prediction of tube temperatures from a SCDAP/RELAP5 analysis. Specific mixing parameter recommendations for SCDAP/RELAP5 analysis of the natural circulation flows are outlined below for the Westinghouse and Combustion Engineering designs.

9.1 Parameter Recommendations for Westinghouse Plant Analysis

The results of these analyses suggest a modification of the mixing parameters used in the analysis of Westinghouse type plants. NUREG-1570 (Ref. 3) suggests best estimate values for the tube split ratio, mixing fraction, and recirculation ratio of 53% hot, 0.87, and 1.9, respectively. These values have been used extensively in the analysis of natural circulation flows during the station blackout transient (TMLB'). New values are suggested after reevaluating the test data and in consideration of these CFD predictions. Suggested values for the tube flow split ratio, mixing fraction, and recirculation ratio for use in Westinghouse plant analysis are given below:

tube split ratio	50% hot flow tubes
mixing fraction	0.81
recirculation ratio	2.7

These values do not come from a single CFD prediction or data set. The estimates provided here are based upon an evaluation of the test data and the CFD predictions completed. Some engineering judgment is used to estimate these values. Background information and other considerations are provided below to give an indication of the rationale used to suggest the values provided above. It is recognized that the basis for these estimates relies on extrapolations and interpretations of a very limited data set. Clearly there is a level of uncertainty in the results that must be considered. Regardless of the uncertainty, it is considered important to provide the author's best estimate of the mixing parameters for use in future system analyses.

The tube flow split ratio is considered first. The 1/7th scale test data was reviewed along with the predictions to estimate this value. An evaluation of the test data indicates a bias in the determination of the percentage of hot tubes. Tubes with no measurement are most often reported as hot flow tubes (only 25% of the tubes had a thermocouple). This bias results in an overstatement of the number of hot flow tubes by as much as 12% in specific cases. The average number of hot flow tubes reported for transient tests SG-T1 through SG-T4 in the 1/7th scale test report is 53%. The bias estimate for these tests is approximately +3%, leading to a suggested percentage of hot flow tubes equal to 50%

The CFD predictions indicated that the percentage of hot tubes is affected by the tube bundle heat transfer rate. The prediction for the model 44 (prototype design) steam generator did not have a tube bundle heat loss rate consistent with the SCDAP/RELAP5 predictions (SCDAP/RELAP5 predictions are assumed to be prototypical). An adjustment to the model 44 predictions is determined by considering the scaleup predictions along with the model 44 results. The CFD predictions for the high-temperature scaleup case indicate 46.3% of the tubes carry hot flow gas for case h1 (see Table 5). Case h1 is considered to have a tube heat loss rate that is most consistent with the SCDAP/RELAP5 predictions. The model 44 predictions indicate 44.4% of the tubes carry hot flow. This value is approximately 2.7% to 6.9% larger than the scaleup predictions (h3 and h4) that have a comparable tube bundle heat loss rate (see Table 7). It is suggested that the model 44 design results in a hot tube ratio that is 4.8% (average) higher than the scaleup design. Using the scaleup prediction with heat loss rates consistent with the SCDAP/RELAP5 predictions (h1) and adding an average differential consistent with the differences noted for the model 44 geometry, the tube split ratio is close to 51% (hot). This value is comparable to the 1/7th scale test data result noted above. A similar argument is applied to the low temperature predictions (l3, l4, and low temperature model 44) and the resulting estimate for the percentage of hot tubes is 45%. In considering the differences between the high and low temperature predictions, the higher temperature predictions are weighted more heavily since most of the heat transfer occurs during the period of rapid core oxidization where the temperatures in the hot leg are close to those used in the high temperature predictions. Obviously there is some uncertainty in the prediction of this value. A value of 50% for the tube split ratio is suggested for future analysis of Westinghouse plants. It is in line with the high temperature estimates and the test data. It is also suggested that values as low as 41% and as high as 51% could be considered in a sensitivity analysis.

The mixing fractions predicted for the model 44 design, 0.80 and 0.81 (see Tables 7 and 8), are lower than the scaleup model predictions. This was discussed earlier and results from the differences in the inlet plenum designs. The scaleup predictions resulted in values ranging from 0.83 to 0.87 (case h1 to h3) for the high temperature predictions and from 0.96 to 0.99 for the low temperature predictions. More mixing is predicted for the scaleup model with its relatively taller inlet plenum design. A consideration of the 1/7th scale transient tests SG-T1 through SG-T4 results in an average mixing fraction of 0.81. This is consistent with the model 44 predictions noted above. Presumably other test data were averaged to obtain the 0.87 value suggested in NUREG-1570. The trends with tube bundle heat transfer rates are less clear for the mixing fraction and no attempt is made to adjust the model 44 predictions. A value of 0.81 is consistent with the predictions for the model 44 design and the transient test data noted above.

The recirculation ratio shows a clear trend with the tube bundle heat transfer rates in the scaleup predictions. An adjustment to the model 44 predictions is estimated to account for the difference between the model 44 predictions heat transfer rate and the assumed prototypical behavior predicted by the SCDAP/RELAP5 code. A look at the available data indicates a range of recirculation ratios that may be considered. The recirculation ratios predicted at full-scale conditions are estimated to be 0.1 higher than the steady-state 1/7th scale test data. This is based upon a comparison of predictions h5 and l5 with a steady-state 1/7th scale test. Transient 1/7th scale tests SG-T1 through SG-T4 have an average recirculation ratio of 2.16. Adding 0.1 from the differences noted above results in a suggested fullscale recirculation ration of 2.26.

The high-temperature scaleup predictions indicate a recirculation ratio in the range from 2.91 to 2.94 (cases h2 and h1) in the assumed prototypical heat transfer range. The value drops to an average value of 2.685 in the range of heat transfer rates consistent with the model 44 predictions. A differential of 0.24 is suggested to account for the difference in the heat transfer rates between the model 44 predictions and the SCDAP/RELAP5 predictions. The high-temperature prediction for the model 44 recirculation ratio, 2.47, is adjusted to 2.71 to account for the estimated effects of the tube bundle heat transfer rate. A similar analysis of the low temperature predictions results in an estimated recirculation ratio of 2.44. A single value for system code analysis of 2.7 is suggested for the recirculation ratio based on a bias towards the high temperature predictions. Values ranging from 2.25 to 2.75 are suggested for sensitivity analysis.

9.2 Parameter Recommendations for Combustion Engineering Plant Analysis

The results of these analyses suggest a modification of the mixing parameters used in the analysis of Combustion Engineering plants. Suggested values for the tube flow split ratio, mixing fraction, and recirculation ratio for use in Westinghouse plant analysis are given below:

tube split ratio   42% hot flow tubes
mixing fraction   0.61
recirculation ratio   1.53

These data are based on two predictions of the flow behavior for a Combustion Engineering plant (see Table 12). The values listed are the average value from the two predictions. Some variation with temperature is shown in the table but the limited number of predictions makes it difficult to draw conclusions about the significance of these variations. The values suggested provide a good starting point for the evaluation of severe accident natural circulation in a Combustion Engineering plant. Sensitivity studies to determine the significance of the uncertainty in these parameters are recommended.

10 REFERENCES

1. Memorandum outlining the steam generator action plan from Ashok Thadani and Samuel Collins to William Travers, dated May 11, 2000.

2. P.D. Bayless et al., "Severe Accident Natural Circulation Studies at the INEL," NUREG/CR-6285, INEL-94/0016, February 1995.

3. SGTR Severe Accident Working Group, "Risk Assessment of Severe Accidnet-Induced Steam Generator Tube Rupture," NUREG-1570, March 1998.

4. C.F. Boyd, K.Hardesty, "CFD Analysis of 1/7th Scale Steam Generator Inlet Plenum Mixing During a PWR Severe Accident," NUREG-1781, October 2003.

5. J.D. Anderson, *Computational Fluid Dynamics: The Basics with Applications*, McGraw-Hill Series in Mechanical Engineering, McGraw Hill, 1995.

6. D.B. Ebeling-Koning et al., "Steam Generator Inlet Plenum Mixing Model for Severe Accident Natural Circulation Conditions," 1990 ASME Winter Meeting, Dallas, Texas.

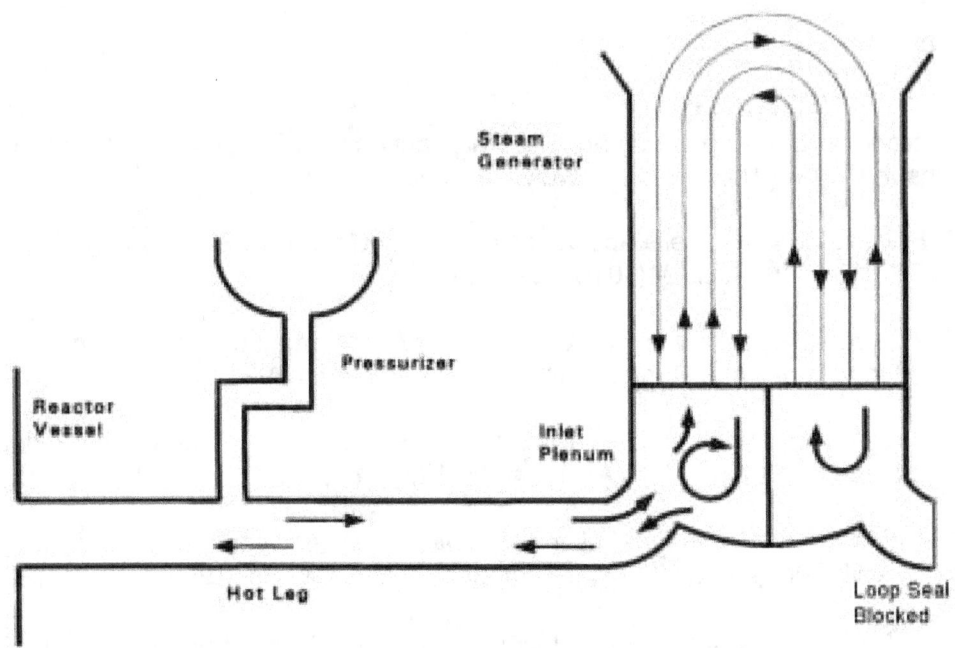

Figure 1. Overview of Natural Circulation Flow Pattern

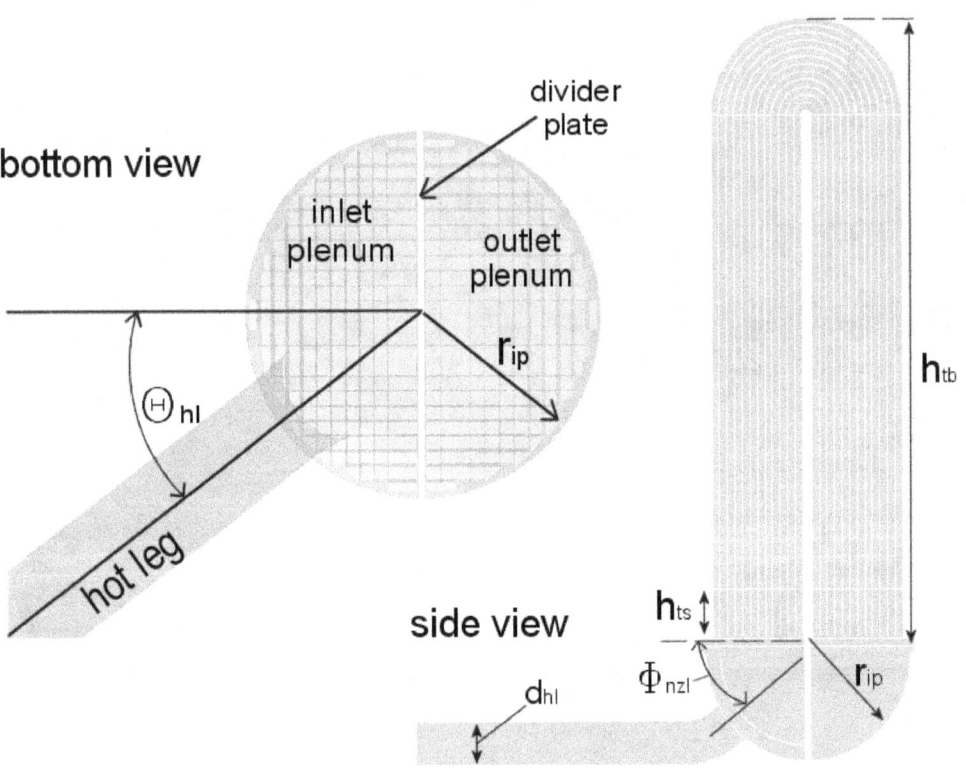

Figure 2. Basic Steam Generator Geometry (Ref. Table 1)

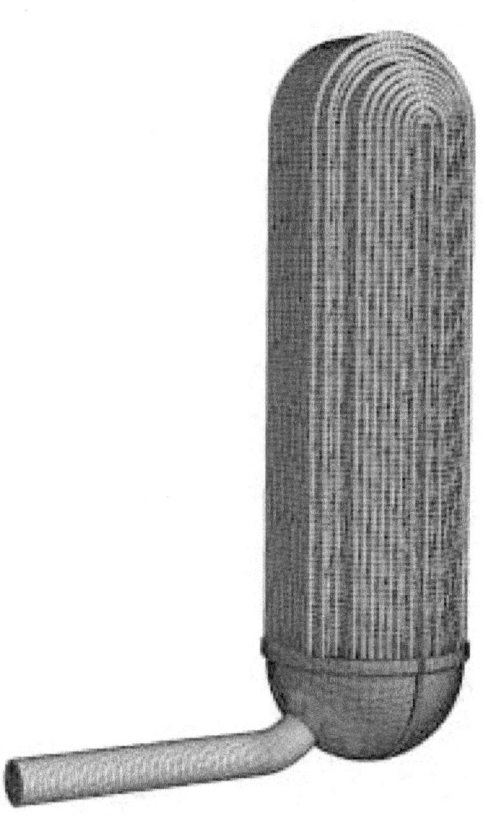

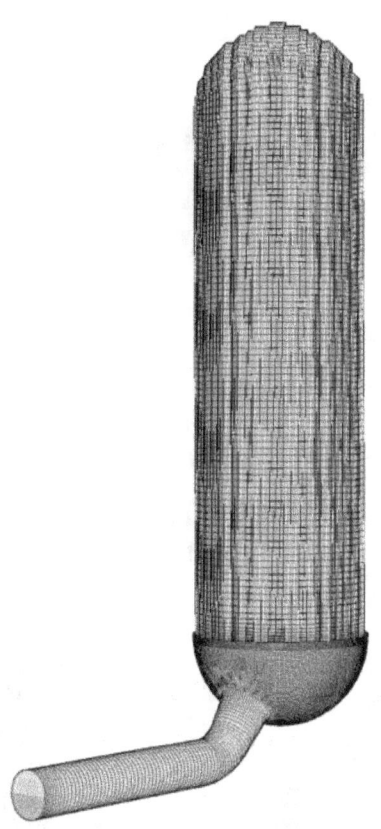

Figure 3. Scale-Up Computational Mesh   Figure 4. Model 44 Computational Mesh

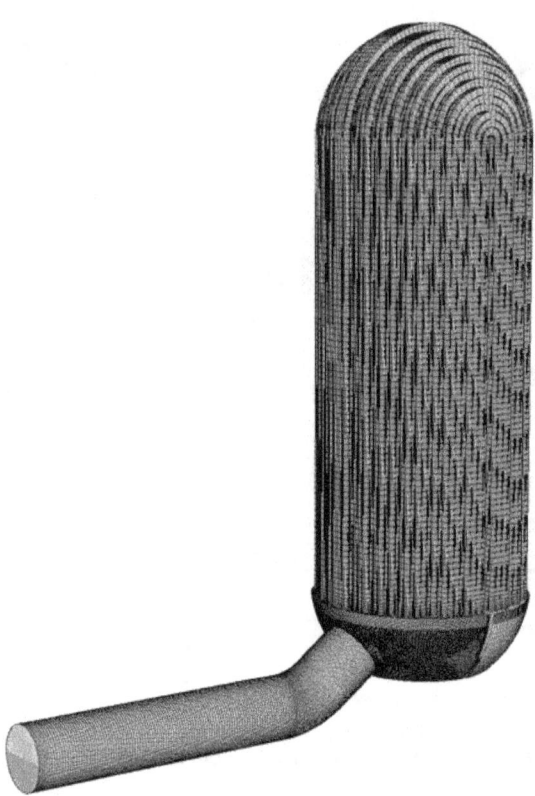

Figure 5. Combustion Engineering Computational Mesh

43

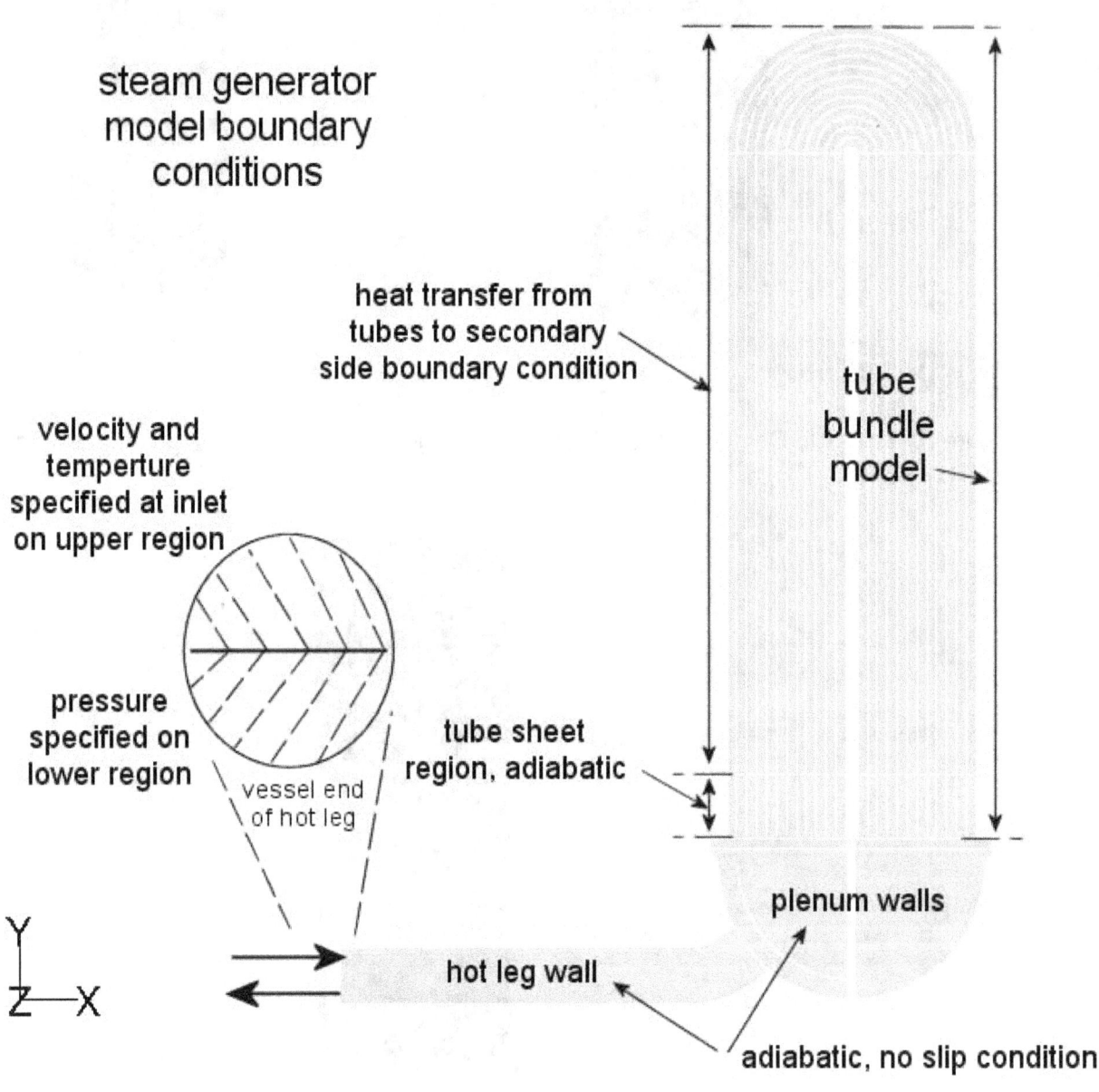

Figure 6. Overview of Model Boundary Conditions

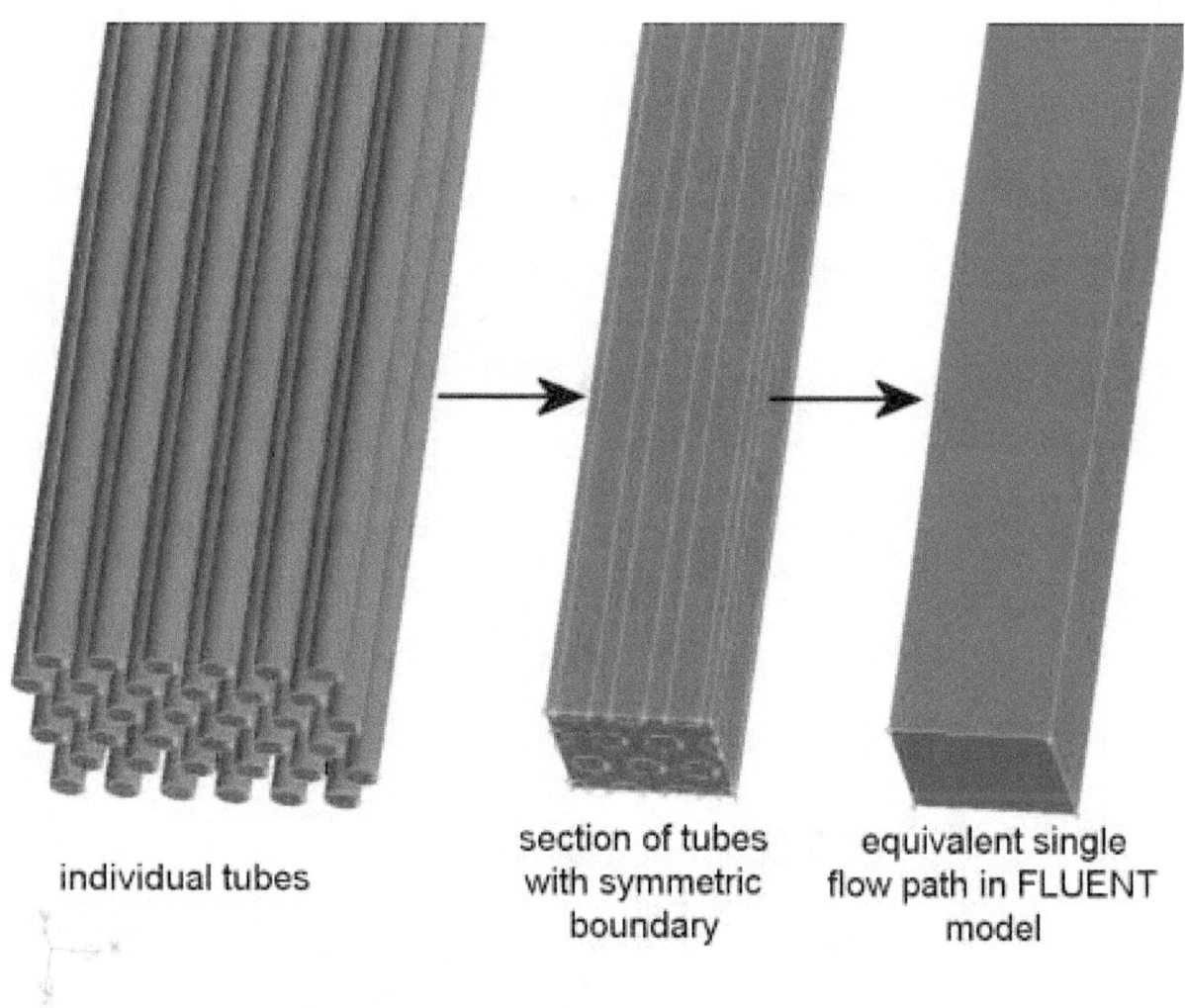

individual tubes   section of tubes with symmetric boundary   equivalent single flow path in FLUENT model

Figure 7. Tube Modeling Approach

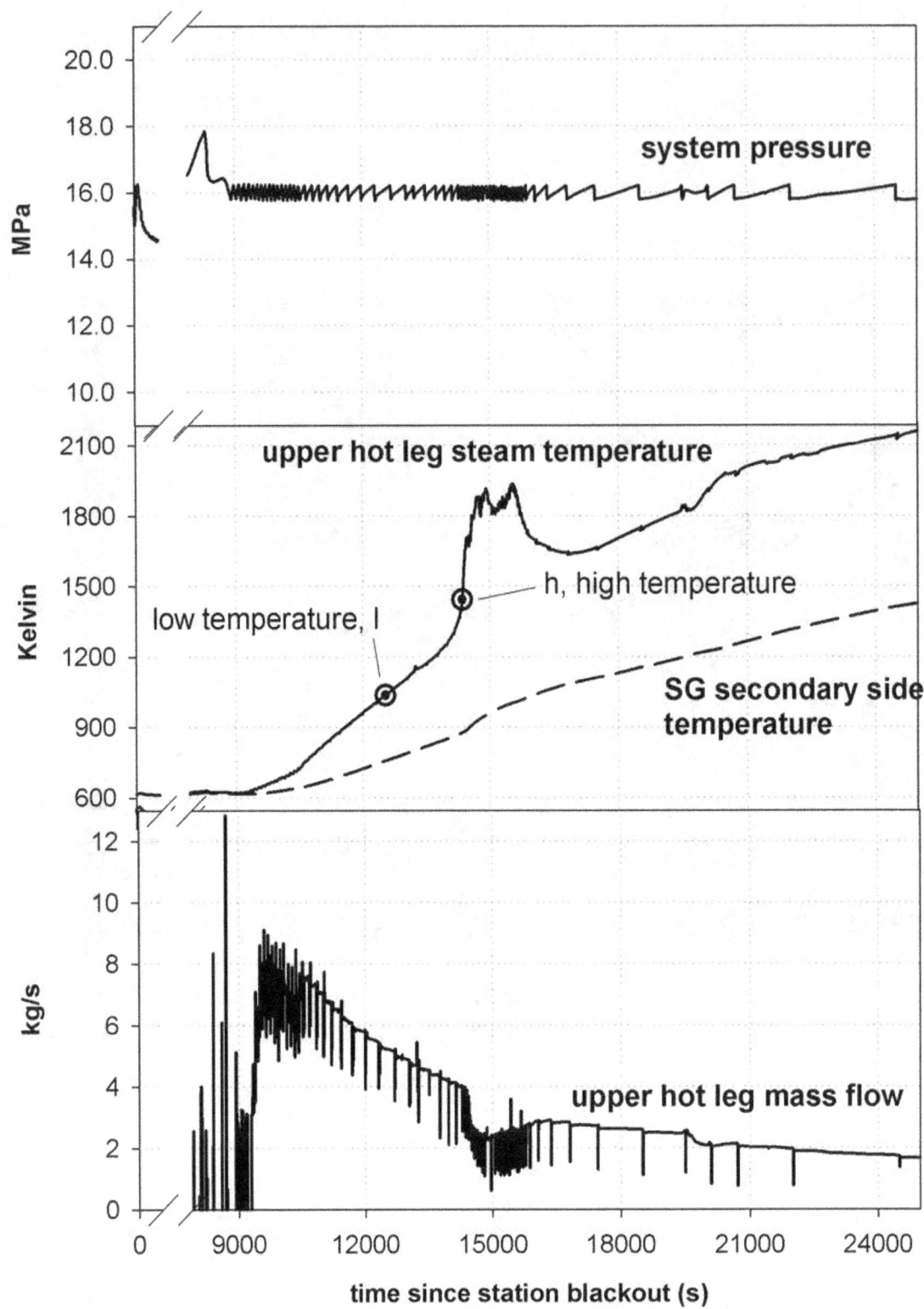

Figure 8. SCDAP/RELAP5 Predictions Used To Select Westinghouse Boundary Conditions

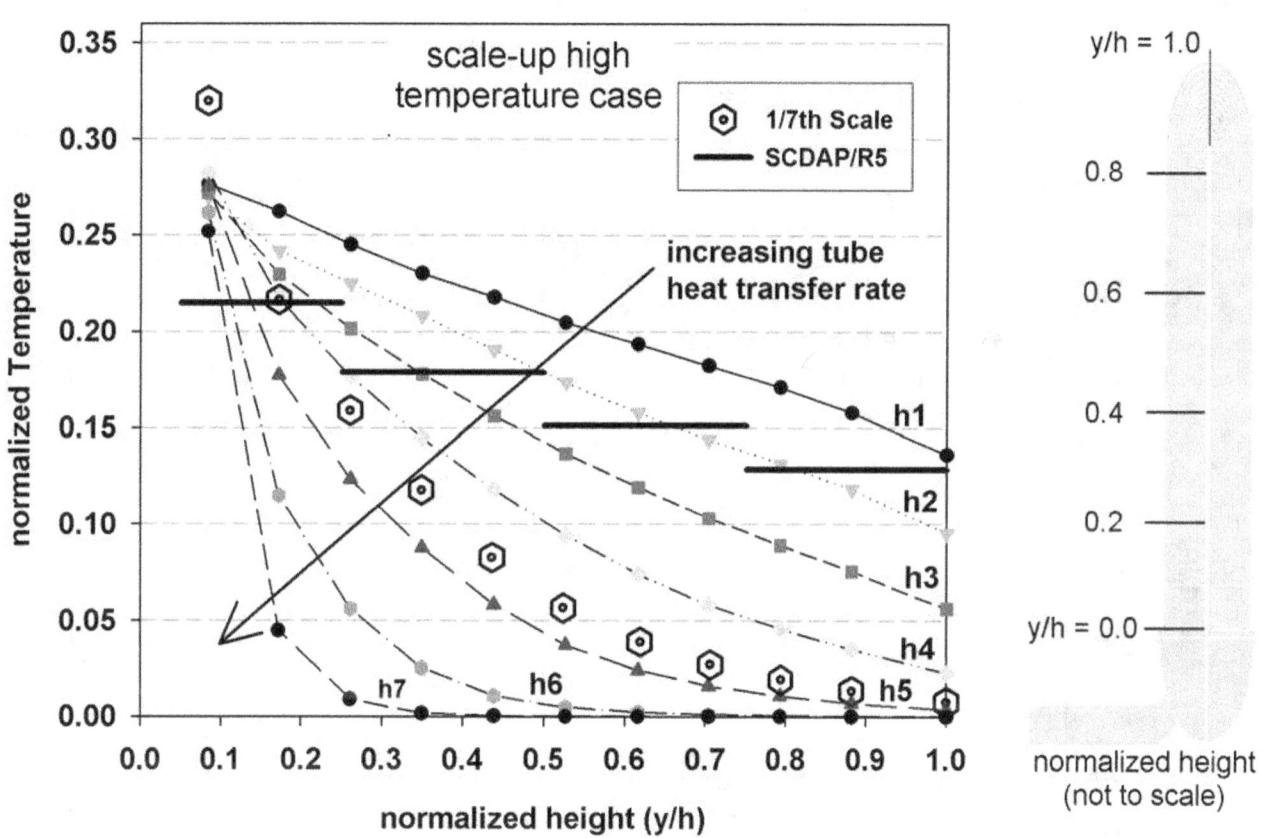

Figure 9. Normalized Temperatures in Tubes for Various Heat Transfer Rates (Scale-Up)

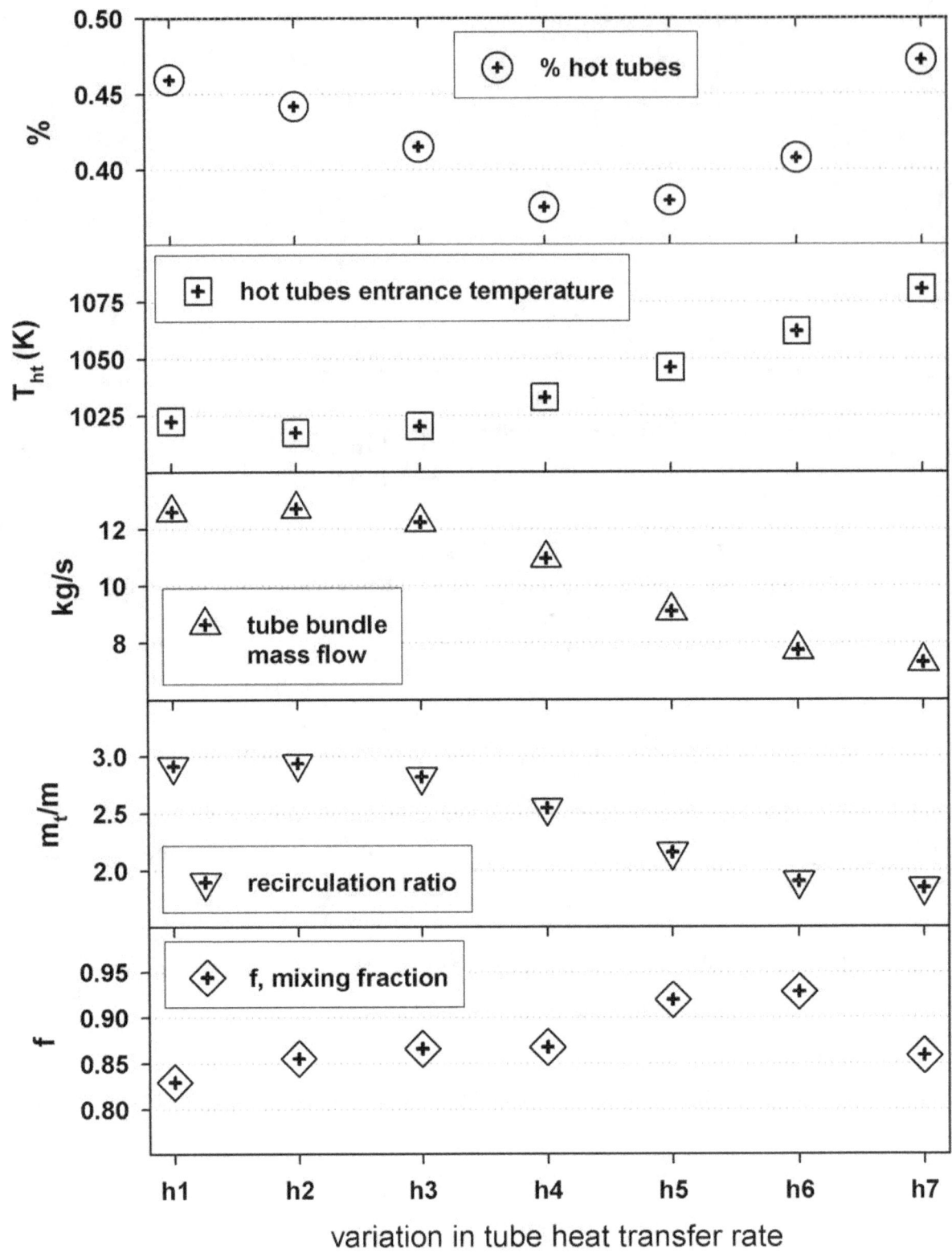

Figure 10. Variations in Mixing Parameters With Tube Heat Transfer Rate

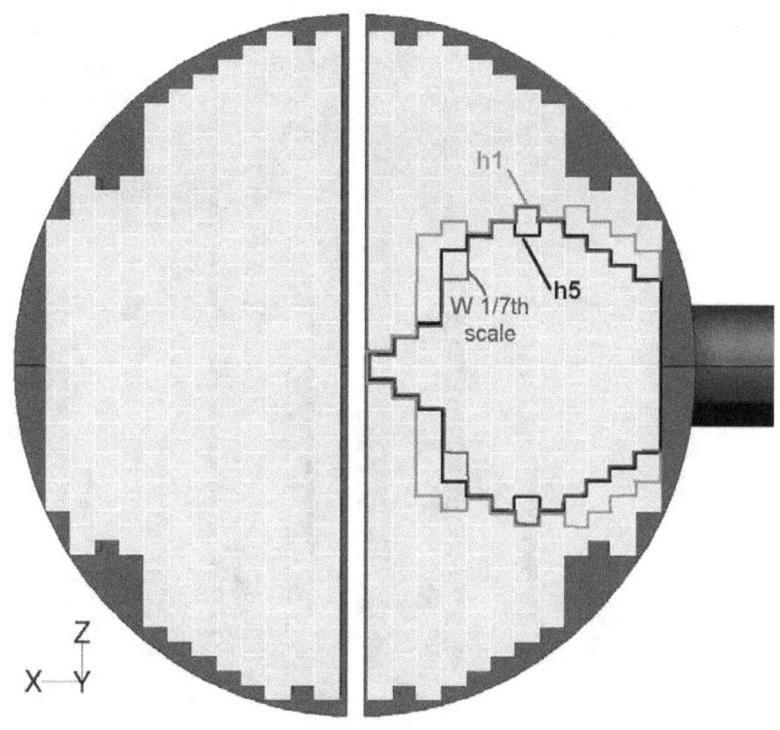

Figure 11. Predicted Boundary Between Hot and Return Flow Tubes

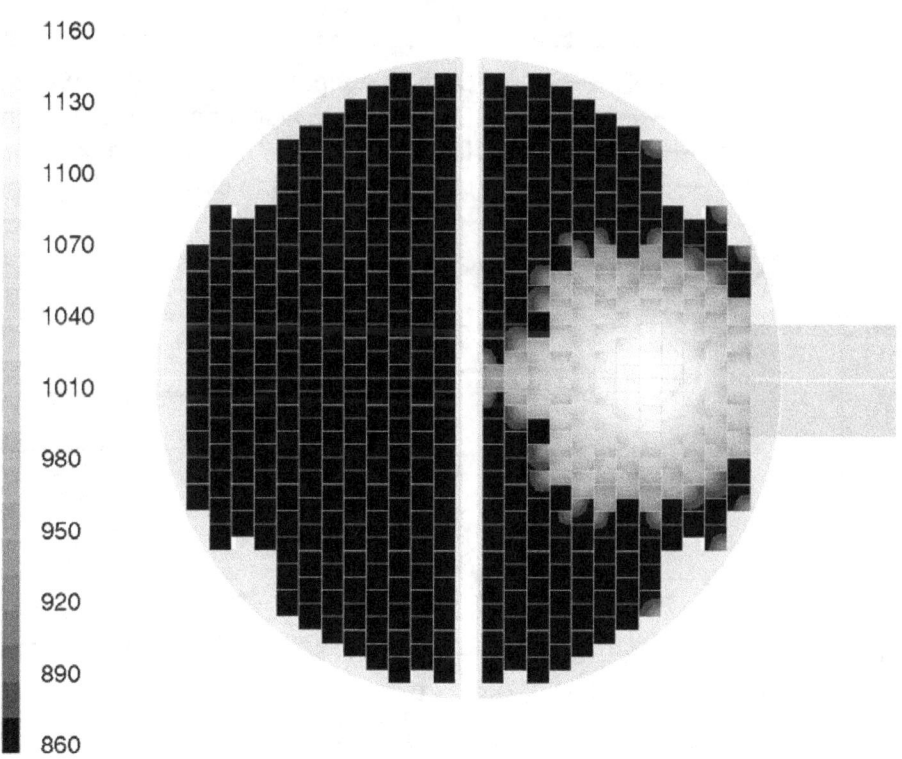

Figure 12.   Predicted Temperature Contours Above Tube Sheet Entrance (case h5)

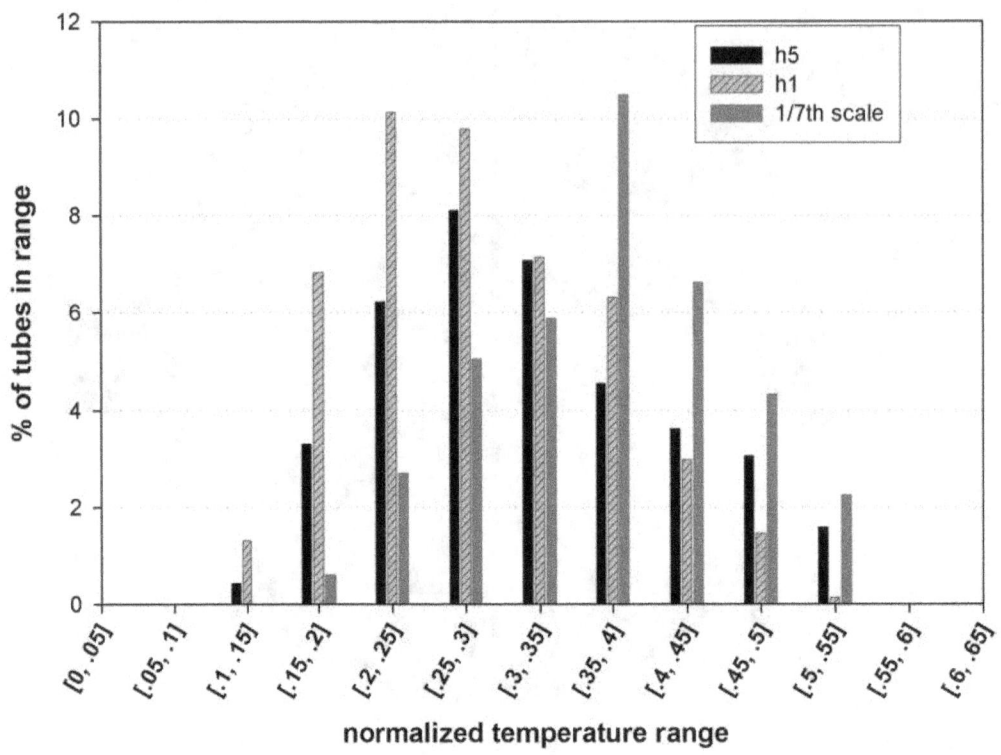

Figure 13. Normalized Temperatures Entering Tube Sheet for Scale-Up Model

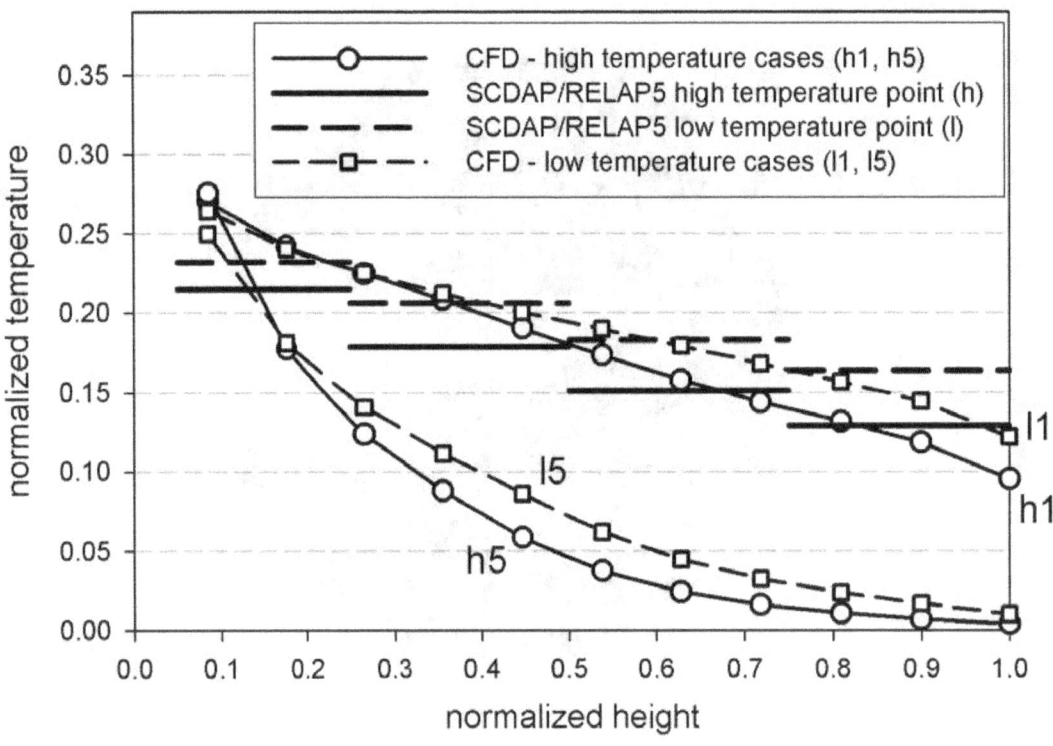

Figure 14. Normalized Average Temperature of Hot Flow Tubes Along Upward Flow Path

50

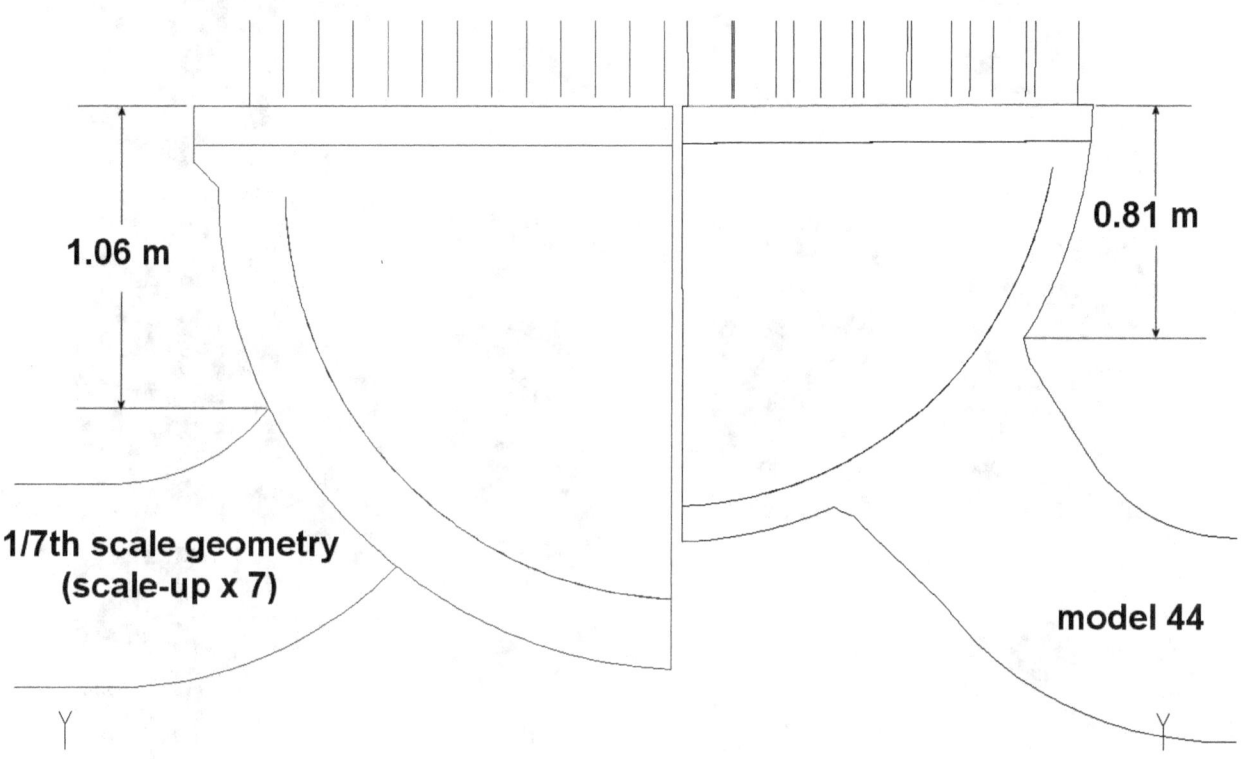

1.06 m

0.81 m

**1/7th scale geometry
(scale-up x 7)**

model 44

Figure 15. Scale-Up Geometry Compared to Prototypical Model 44 Design

51

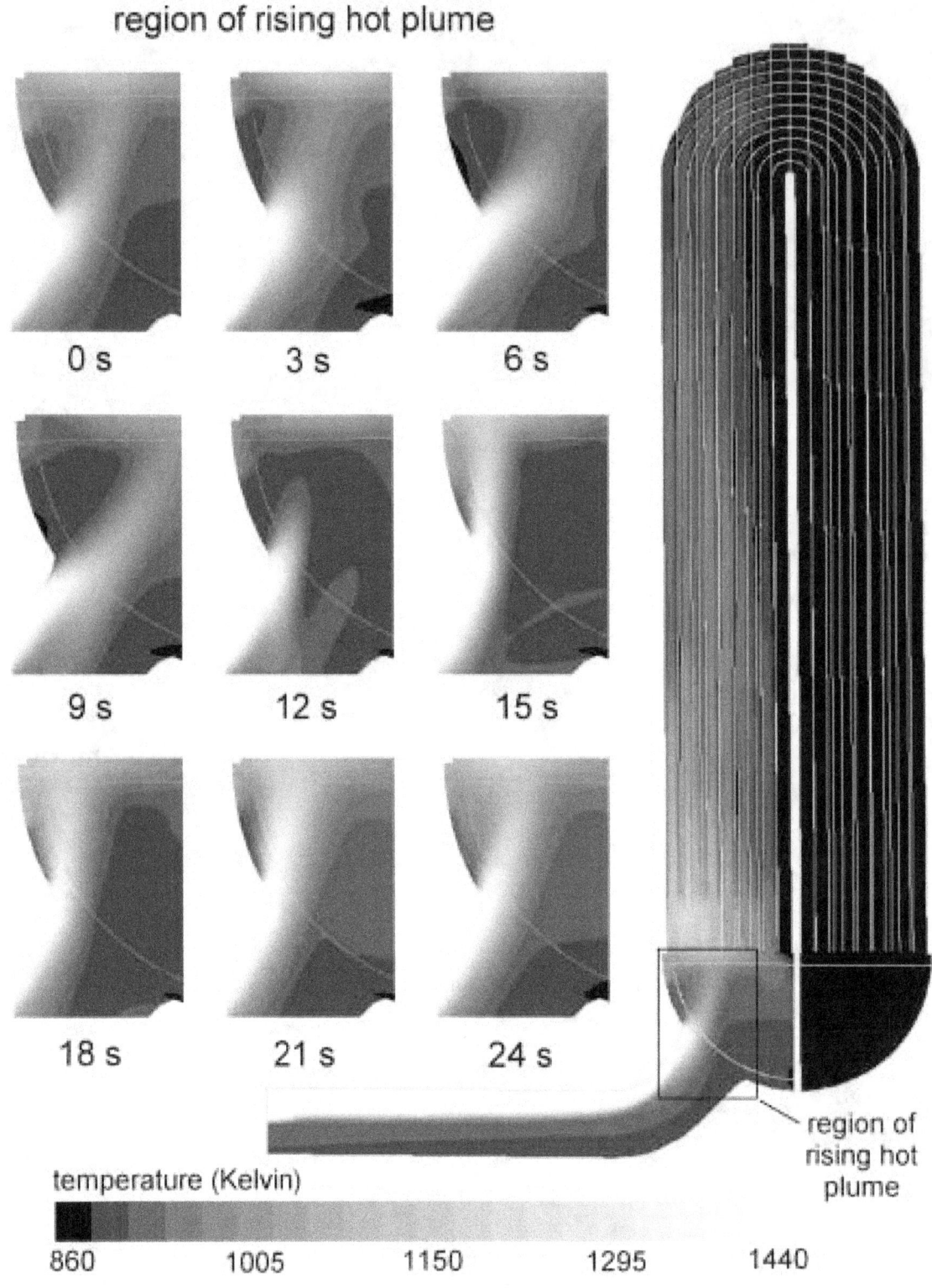

temperature contours for
region of rising hot plume

0 s   3 s   6 s

9 s   12 s   15 s

18 s   21 s   24 s

region of
rising hot
plume

temperature (Kelvin)

860   1005   1150   1295   1440

Figure 16. Temperature Contours for Hot Plume on Hot Leg Symmetry Plane (Model 44)

52

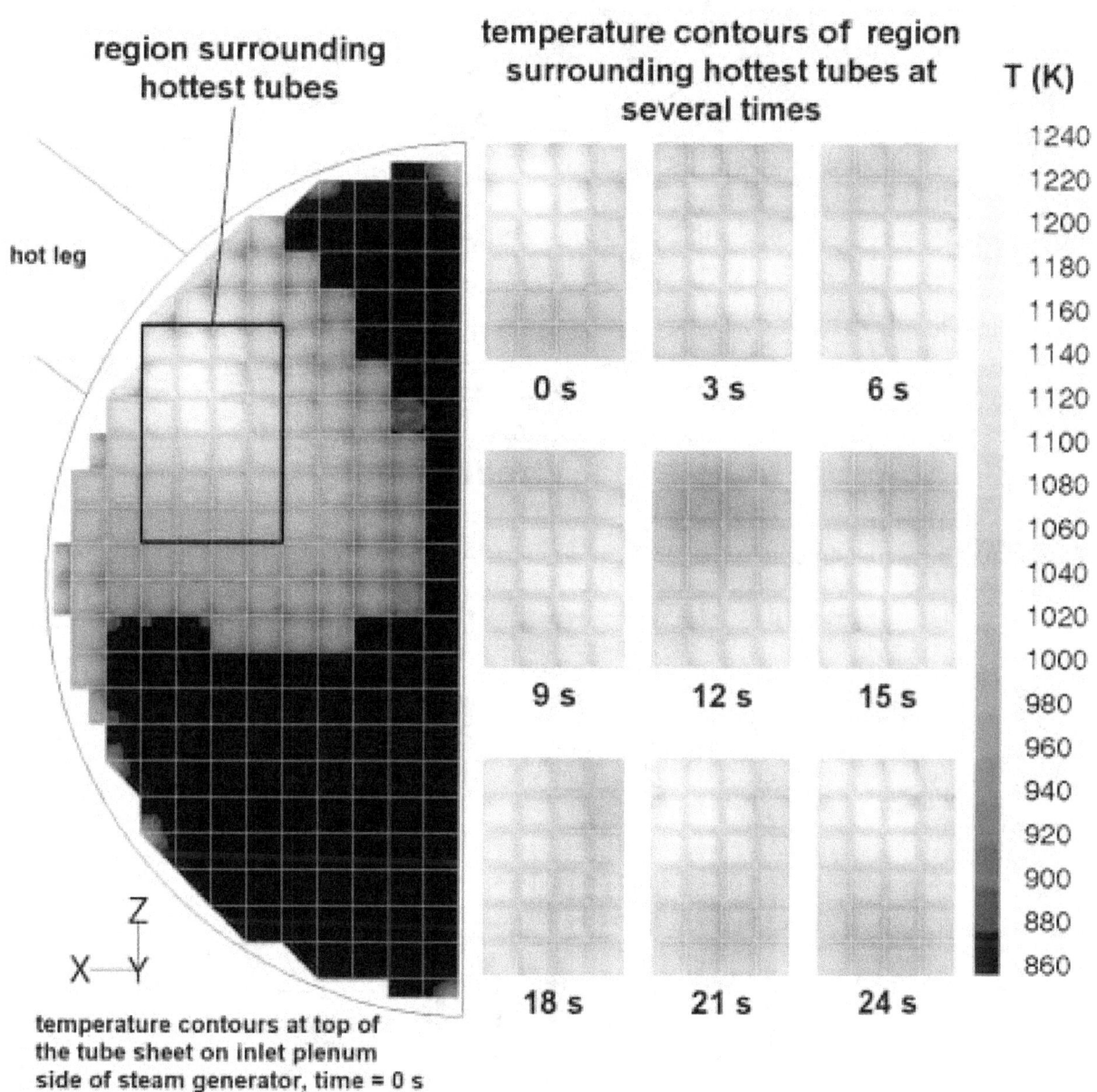

Figure 17. Temperature Contours Just Above Tube Sheet Entrance (Model 44)

53

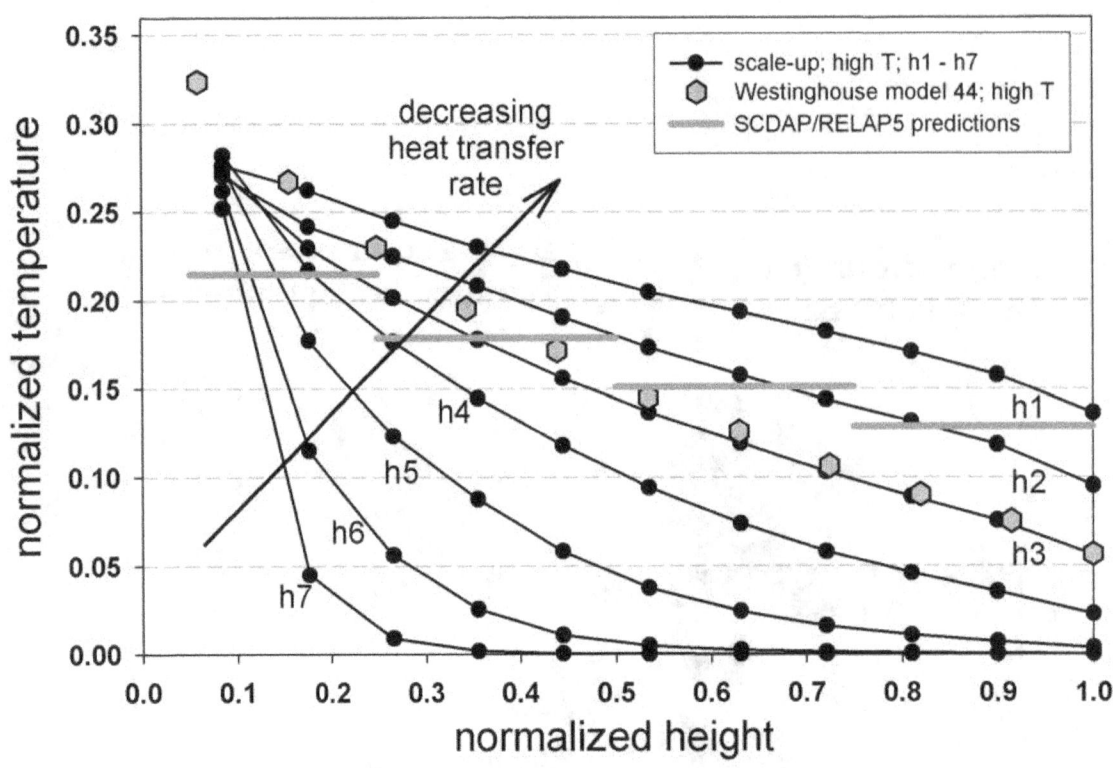

Figure 18. Model 44 and Scale-Up Temperatures Along Hot Tubes (High Temperature)

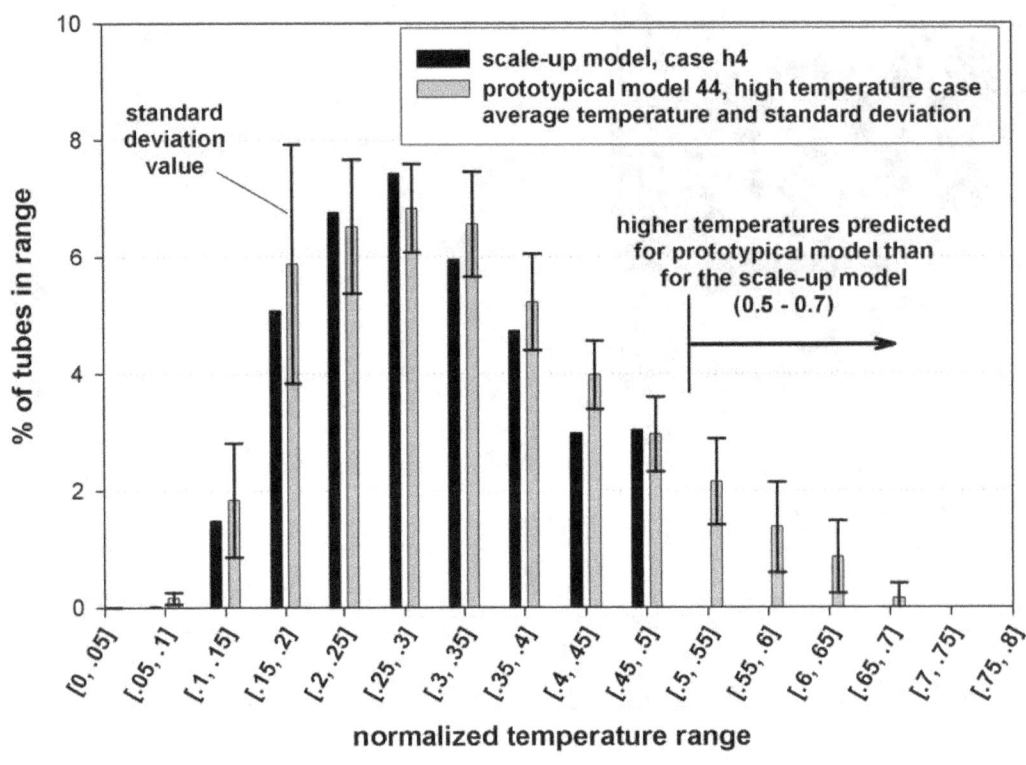

Figure 19. Model 44 and Scale-Up Tube Entrance Temperatures (High Temperature)

54

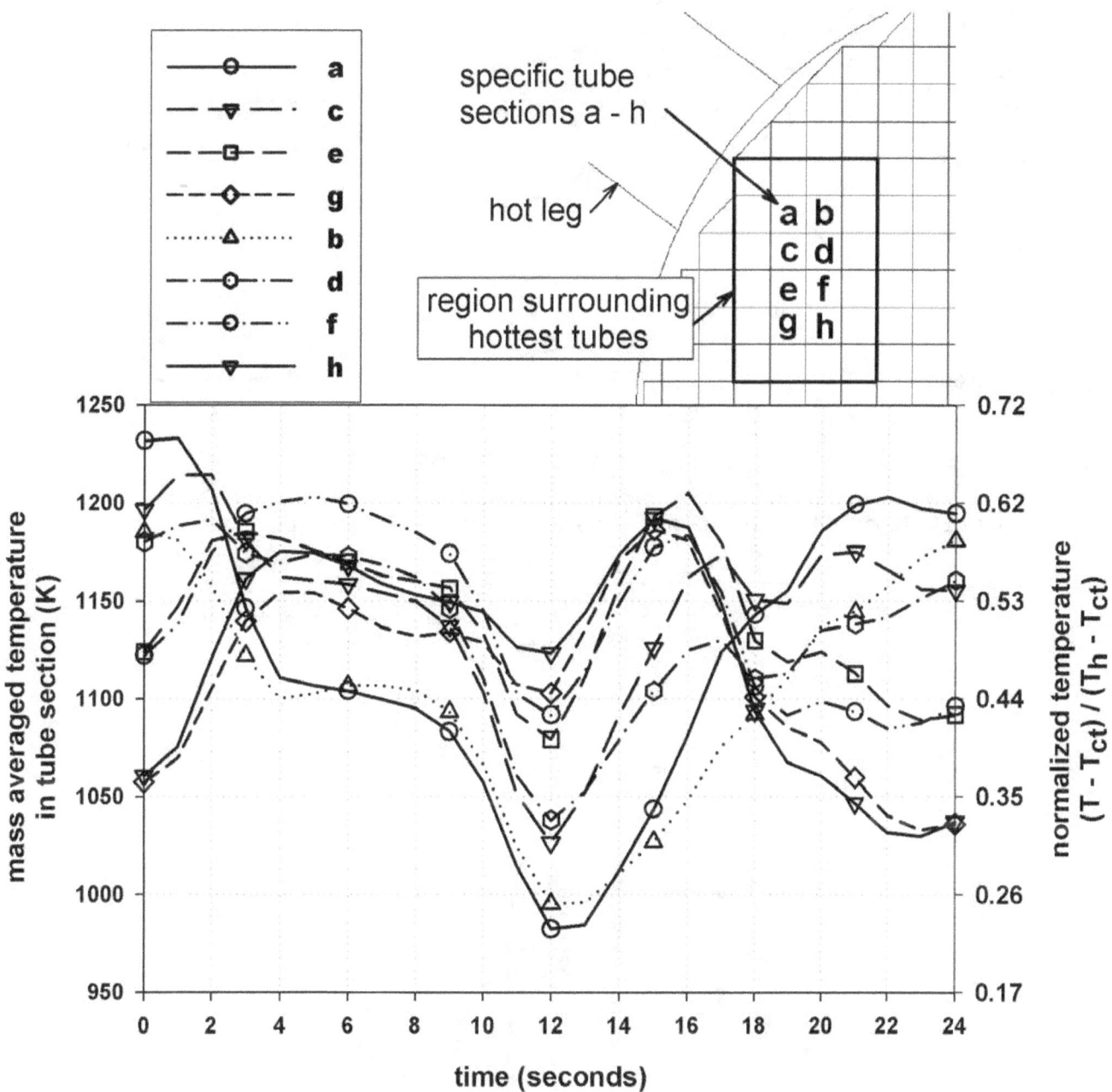

Figure 20. Hottest Tube Regions Variation With Time (Model 44/High Temperature Case)

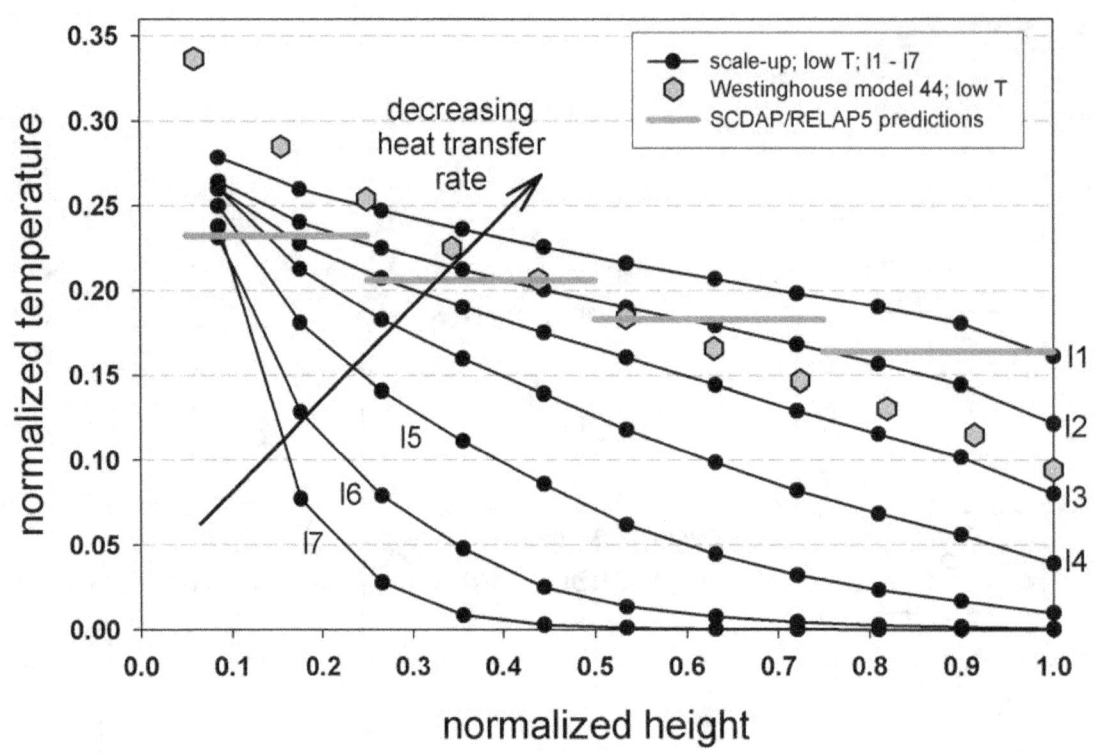

Figure 21. Model 44 and Scale-Up Temperatures Along Hot Tubes (Low Temperature)

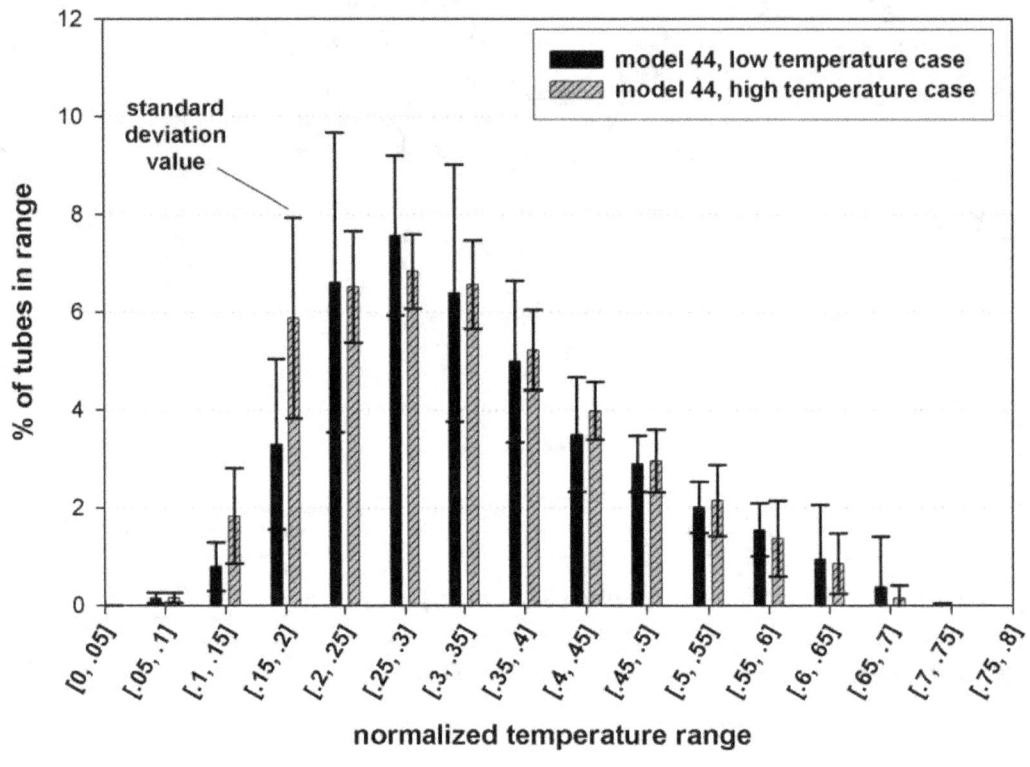

Figure 22. Model 44 Tube Entrance Temperatures for High and Low Temperature Case

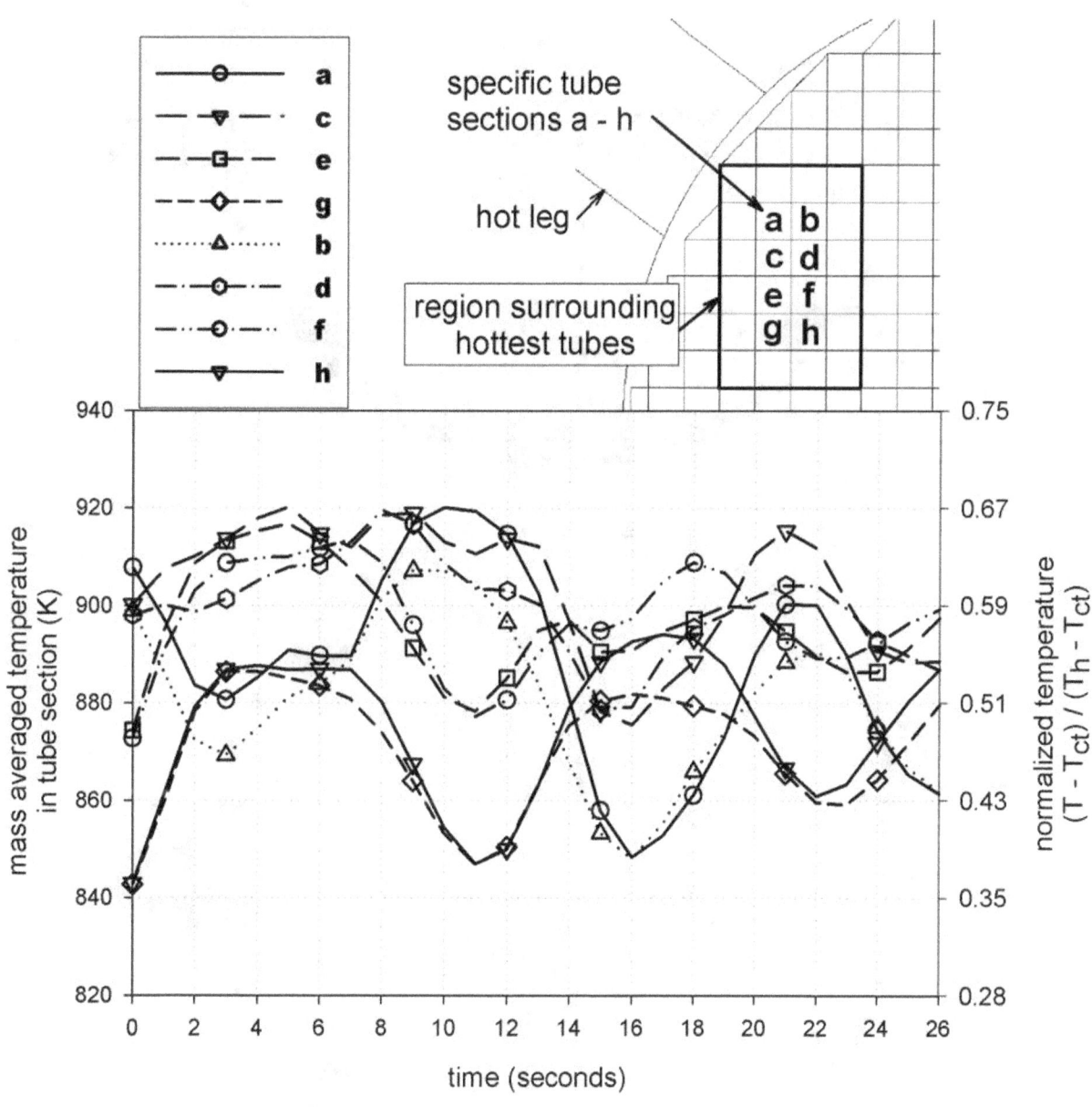

Figure 23. Hottest Tube Regions Variation With Time (Model 44 /Low Temperature Case)

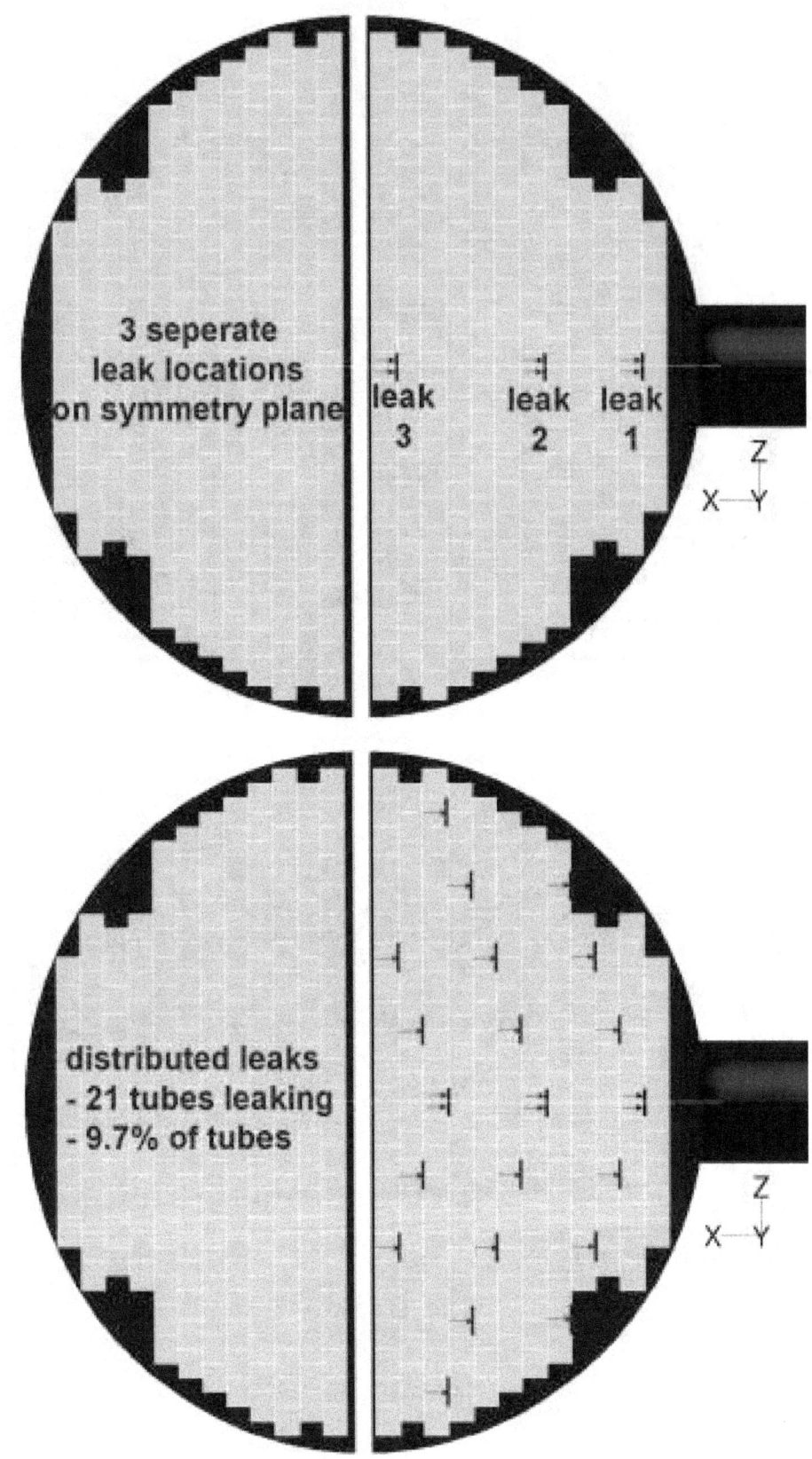

Figure 24. Tube Leakage Locations

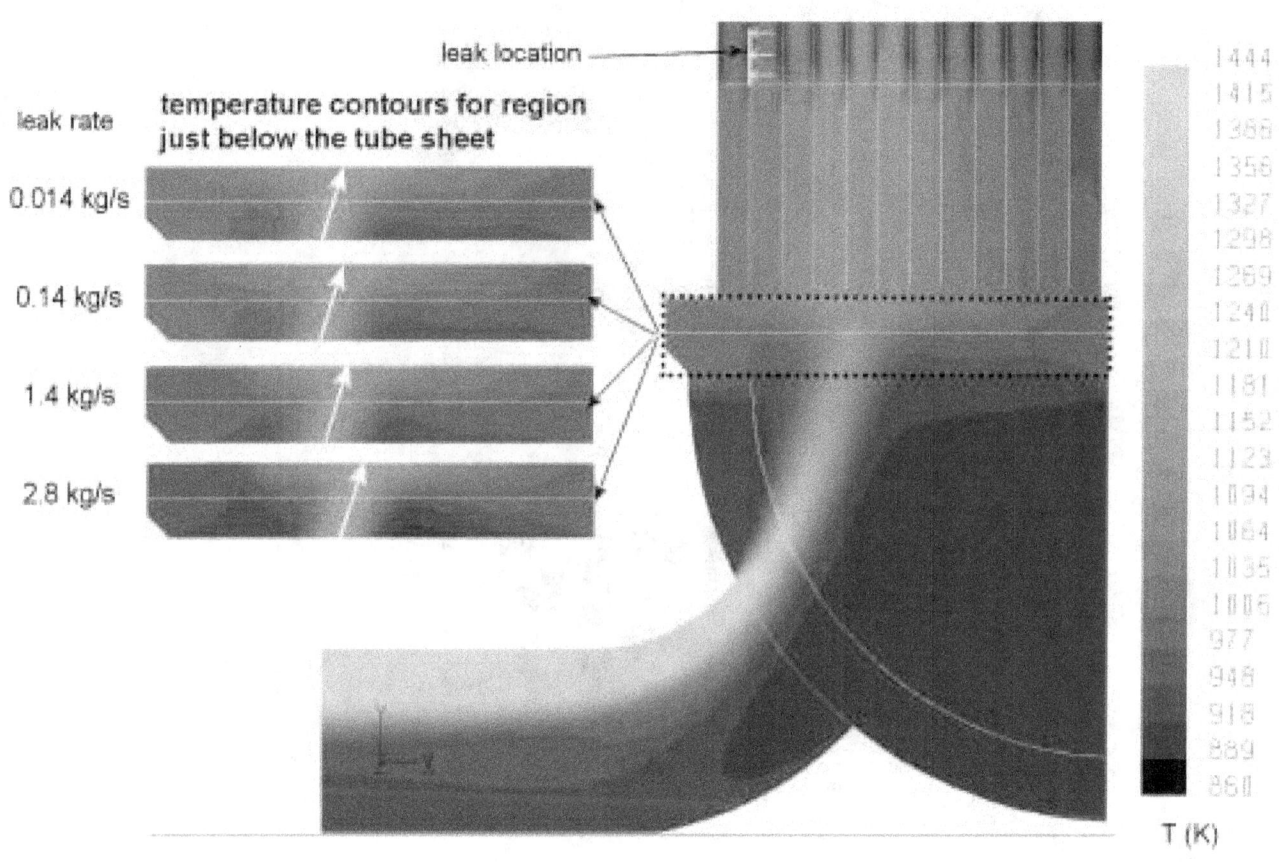

Figure 25. Temperature Contours Near Tube Sheet Entrance for Leak 1

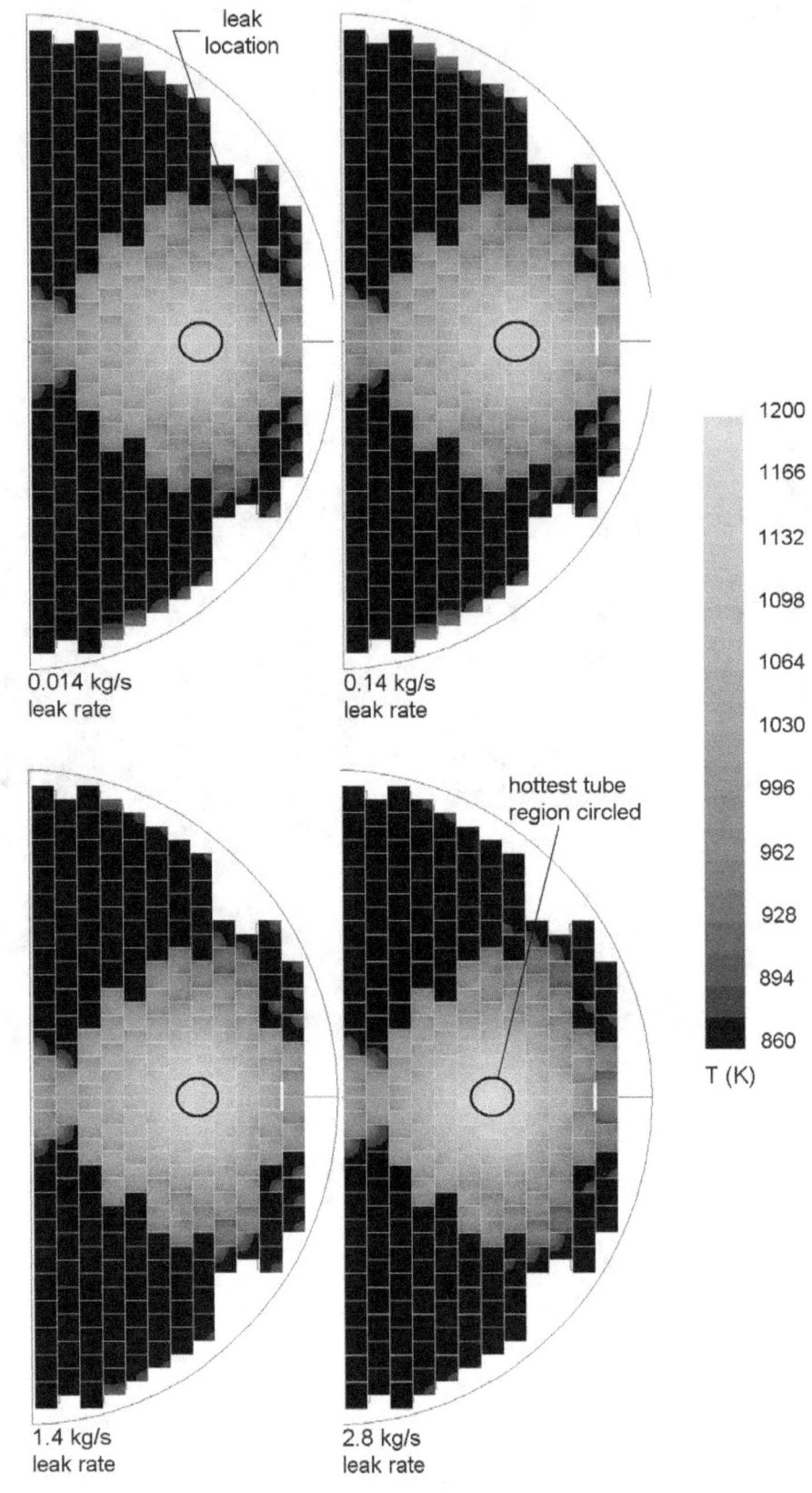

Figure 26. Temperature Contours Above Tube Sheet Entrance for Leak 1

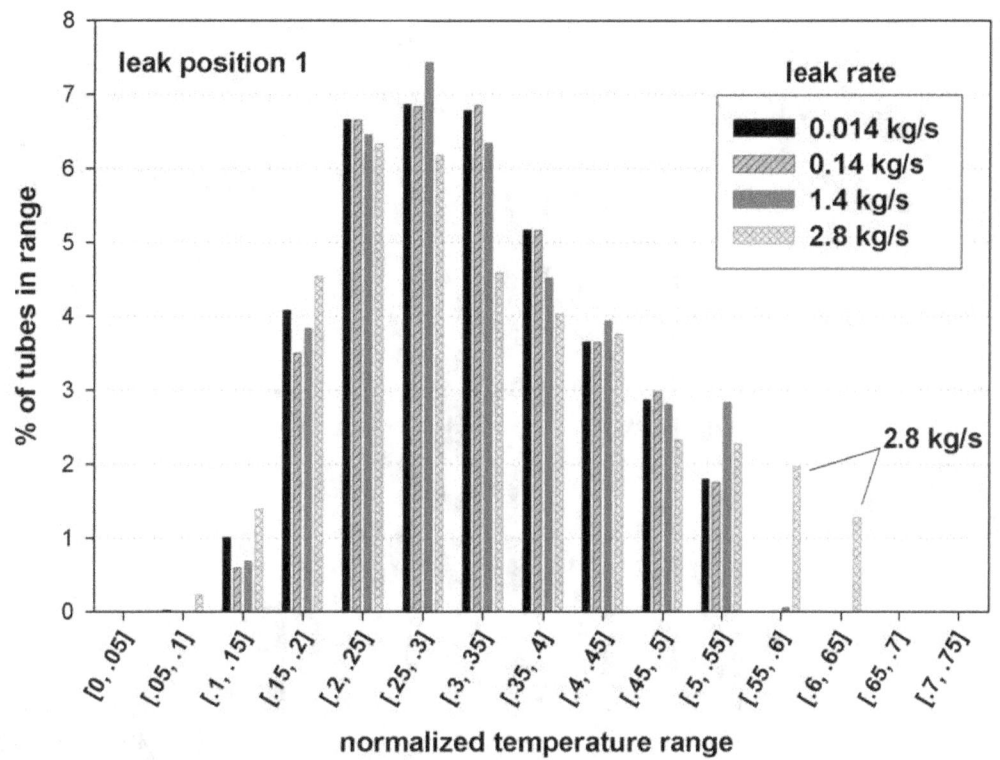

Figure 27. Temperature Histograms for Leak 1

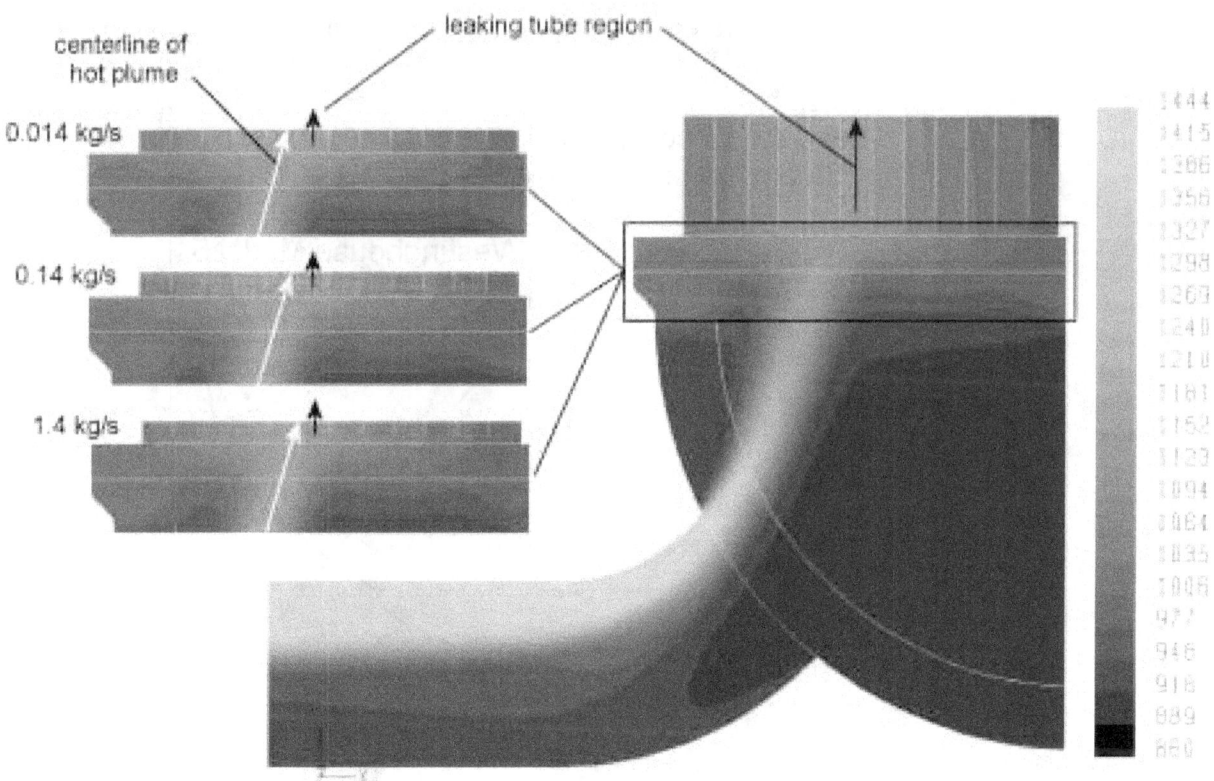

Figure 28. Temperature Contours for Leak Position 2

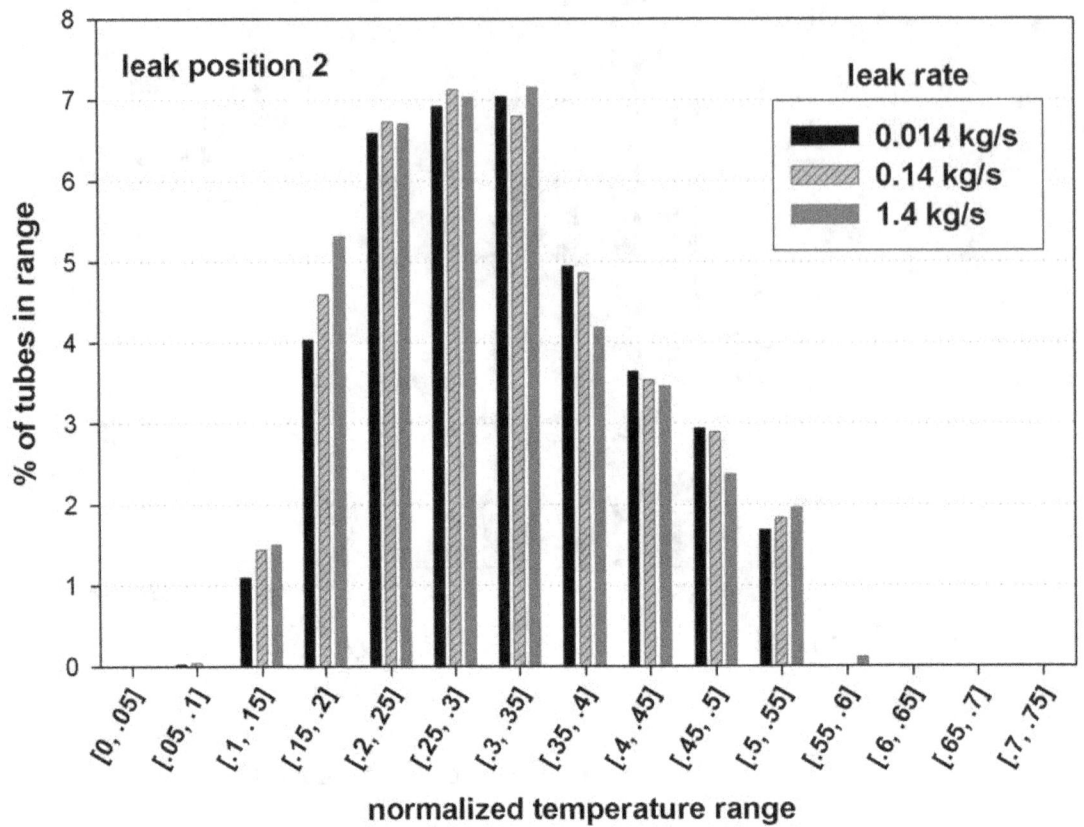

Figure 29. Temperature Histograms for Leak Position 2

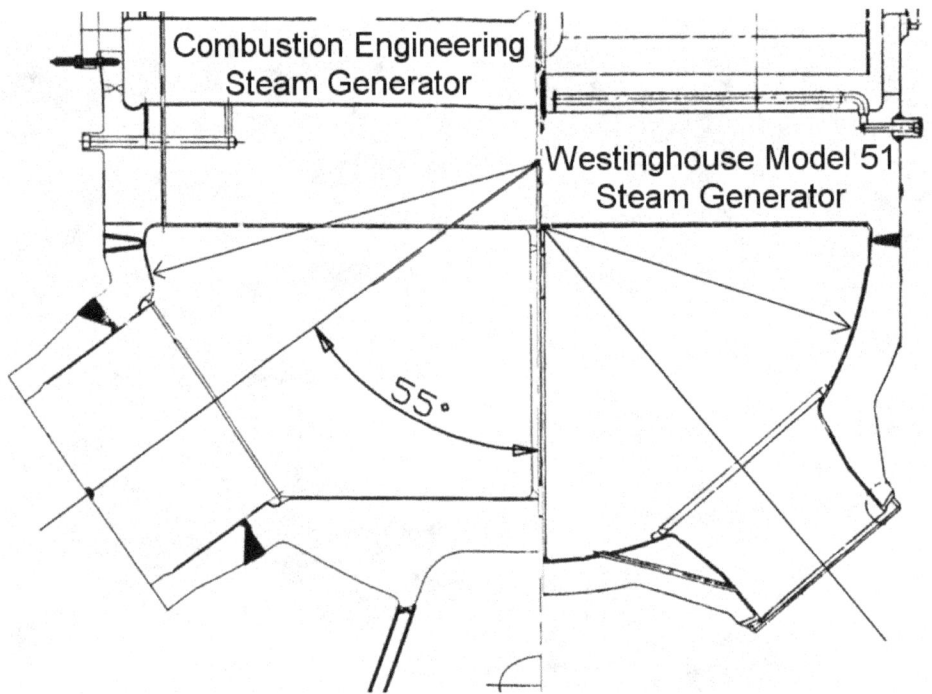

Figure 30. Comparison of CE and Westinghouse Geometry

1315   T (Kelvin)

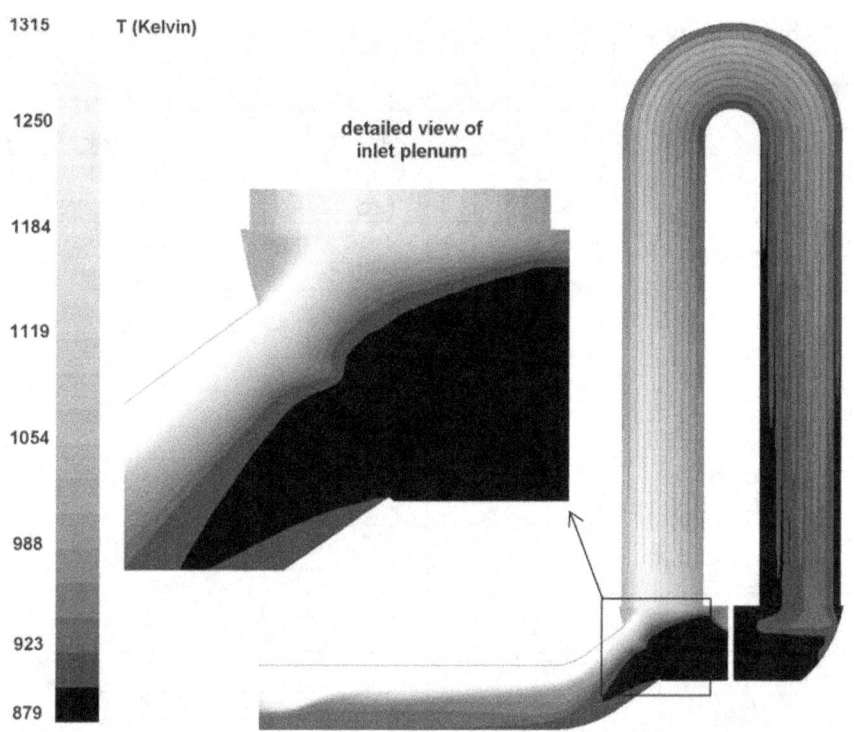

Figure 31. Temperature Contours on Symmetry Plane of CE Model

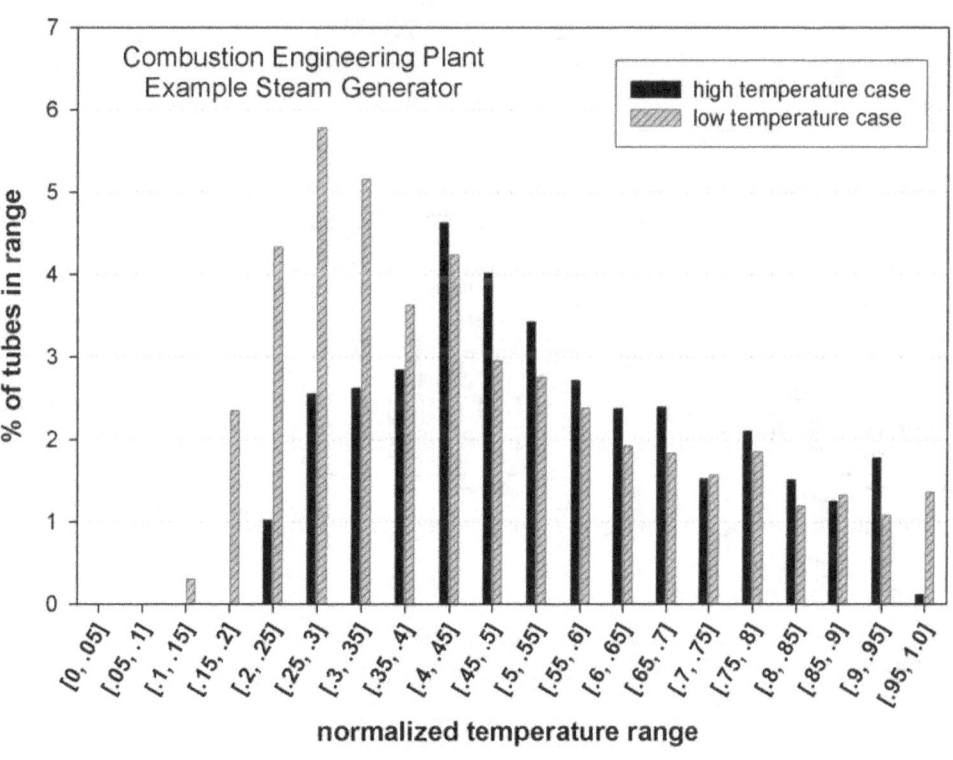

Figure 32. Predicted Tube Entrance Temperatures for CE Model

63

Table 1. Overall Dimensions of Steam Generators (As Modeled)

Dimension	Westinghouse 1/7th facility	Scale-up Model	Westinghouse Model 44	CE Plant Generator
hot leg inner diameter, d_{hl} m (inches)	0.1022 m (4.03")	0.7158 m (28.18")	0.7366 m (29")	1.0668 m (42")
inlet plenum radius, r_{ip} m (inches)	0.2413 m (9.5")	1.689 m (66.5")	1.508 m (59.38")	1.9685 m (77.5")
hot leg orientation, Q_{hl} degrees	0	0	36.5	0
nozzle angle, F_{nzl} degrees	45	45	50	35
total number of U-tubes	216	3216	3216	8741
tube inner diameter, d_t m (inches)	0.00775 m (0.305")	0.01968 m (0.775")	0.01968 m (0.775")	
thickness of tube sheet, h_{ts} m (inches)	0.1143 m (4.5")	0.8001 m (31.5")	0.5578 m (21.96")	0.5652 m (22.25")
height of tube bundle from bottom of tube sheet, h_{tb} m (inches)	1.43 m (56.3")	10.01 m (394.1")	10.04 m (395.3")	
tube array tube pitch m (inches)	triangular 0.02064 m (0.8125")	square 0.03135 m (1.2344")	square 0.03135 m (1.2344")	triangular 0.0254m (1.0")
Total number of tube flow paths in CFD model	216	216	201	282

Table 2. Mesh Characteristics

Description	Scale-up Model	Westinghouse Model 44	CE Plant Generator
number of computational cells for full 3D model (symmetric models use 50% of value in table)	1,014,746	972,705	2,590,476
symmetric model	Yes	No	Yes
average number of cells across the hot leg diameter	26	26	50
average number of cells across the inlet plenum radius	52	50	75
number of individual tube flow paths	216	201	282
average cell dimension (m)	.045	.042	.034

Table 3. Boundary Conditions from SCDAP/RELAP5 (used for Westinghouse Models)

condition	high temperature case (h)	low temperature case (l)
hot leg inlet mass flow, kg/s (lbm/s)	4.09 kg/s (9.01 lbm/s)	5.24 kg/s (11.55 lbm/s)
hot leg inlet temperature, K (^{o}F)	1444 K (2140 ^{o}F)	1024 K (1384 ^{o}F)
T_{ct}, secondary side (sink) temperature (outside tube bundle) K (^{o}F)	860 K (1088 ^{o}F)	750 K (890 ^{o}F)
secondary side heat transfer rate	range of conditions low (h1) to high (h7)	range of conditions low (l1) to high (l7)

Table 4. Scaling Parameters (1/7th Scale Tests and Full-Scale Westinghouse Conditions)

parameter	high temperature case h	low temperature case l	1/7th Scale Data
hot leg Reynolds number Re_{hl}	2×10^{5}	3×10^{5}	6×10^{4}
Inlet hot plume Grashof number Gr_{p}	6×10^{12}	1×10^{13}	8×10^{11}
Richardson number Gr_{p} / Re_{hl}^{2}	2×10^{2}	1×10^{2}	2×10^{2}

Table 5. Scaleup Results/High Temperature Cases h1-h7

Result	h1	h2	h3	h4	h5	h6	h7	1/7th scale prediction
	low tube heat transfer > increasing > high tube heat transfer							
tube heat loss, kW (BTU/s)	5356 (5080)	5426 (5146)	5400 (5122)	5310 (5036)	5047 (4787)	4678 (4437)	4432 (4204)	3.69 (3.50)
% tubes carrying hot gas	46.3	44.4	41.7	37.5	38.0	40.7	47.2	38.0
T_h, hot leg hot temperature, K (°F)	1403 (2066)	1402 (2064)	1400 (2060)	1402 (2064)	1404 (2068)	1407 (2073)	1410 (2078)	428 (311)
T_c, hot leg cold temp., K (°F)	925 (1205)	918 (1193)	919 (1195)	924 (1204)	942 (1236)	964 (1276)	979 (1303)	353 (176)
m, hot leg mass flow, kg/s (lbm/s)	4.3 (9.5)	4.3 (9.5)	4.3 (9.6)	4.3 (9.5)	4.2 (9.3)	4.1 (9.0)	4.0 (8.7)	0.059 (0.129)
T_{ht}, hot tubes temperature, K (°F)	1022 (1380)	1018 (1373)	1020 (1376)	1033 (1400)	1046 (1423)	1062 (1452)	1080 (1484)	373 (212)
m_t, tube bundle mass flow, kg/s (lbm/s)	12.6 (27.8)	12.7 (28.0)	12.2 (26.9)	10.9 (24.0)	9.1 (20.1)	7.7 (17.1)	7.3 (16.1)	0.12 (0.266)
m_t/m, recirculation ratio	2.91	2.94	2.82	2.55	2.16	1.91	1.85	2.06
f, mixing fraction	.83	.86	.87	.87	.92	.93	.86	.81

66

Table 6. Scaleup Results/Low Temperature Cases I1-I7

Result	I1	I2	I3	I4	I5	I6	I7	1/7th scale prediction
	low tube heat transfer > increasing > high tube heat transfer							
tube heat loss, kW (BTU/s)	3340 (3168)	3340 (3168)	3376 (3202)	3317 (3146)	3155 (2992)	2899 (2750)	2701 (2562)	3.69 (3.50)
% tubes carrying hot gas	44.1	44.1	41.4	39.3	39.4	42.6	42.6	38.0
T_h, hot leg hot temperature, K (°F)	1003 (1346)	1003 (1346)	1002 (1345)	1002 (1345)	1005 (1349)	1008 (1355)	1010 (1358)	428 (311)
T_c, hot leg cold temp., K (°F)	778 (941)	778 (941)	776 (937)	778 (941)	785 (953)	795 (971)	803 (986)	353 (176)
m, hot leg mass flow, kg/s (lbm/s)	5.5 (12.2)	5.5 (12.1)	5.5 (12.2)	5.5 (12.1)	5.3 (11.7)	5.1 (11.2)	4.9 (10.8)	0.059 (0.129)
T_{ht}, hot tubes temperature, K (°F)	823 (1021)	823 (1021)	822 (1020)	824 (1024)	829 (1032)	834 (1041)	847 (1065)	373 (212)
m_t, tube bundle mass flow, kg/s (lbm/s)	14.8 (32.6)	14.8 (32.6)	14.5 (31.9)	13.3 (29.3)	11.4 (25.2)	9.6 (21.1)	8.8 (19.3)	0.12 (0.266)
m_t/m, recirculation ratio	2.68	2.69	2.62	2.43	2.15	1.89	1.80	2.06
f, mixing fraction	.94	.94	.96	.99	1.03	1.06	.96	.81

67

Table 7. Results for Prototypical Westinghouse Design at High Temperature Conditions

prediction	prototypical model 44 predictions high temperature conditions		scale-up predictions	
	average value	standard deviation	h3	h4
tube heat loss, kW (BTU/s)	5515 (5227)	52 (49)	5400 (5122)	5310 (5036)
% tubes carrying hot gas	44.4	0	41.7	37.5
T_h, hot leg hot temperature, K (°F)	1404 (2068)	3 (6)	1400 (2060)	1402 (2064)
T_c, hot leg cold temp., K (°F)	924 (1203)	6 (11)	919 (1195)	924 (1204)
m, hot leg mass flow, kg/s (lbm/s)	4.3 (9.6)	0.15 (0.33)	4.3 (9.6)	4.3 (9.5)
T_{ht}, hot tubes temperature, K (°F)	1049 (1428)	12 (22)	1020 (1376)	1033 (1400)
m_t, tube bundle mass flow, kg/s (lbm/s)	10.7 (23.7)	.03 (.07)	12.2 (26.9)	10.9 (24.0)
m_t/m, recirculation ratio	2.47	.09	2.82	2.55
f, mixing fraction	.80	.06	.87	.87

Table 8. Results for Prototypical Westinghouse Design at Low Temperature Conditions

prediction	prototypical model 44 predictions low temperature conditions		scale-up predictions	
	average value	standard deviation	I3	I4
tube heat loss, kW (BTU/s)	3525 (3341)	16 (15)	3376 (3202)	3317 (3146)
% tubes carrying hot gas	41.2	0	41.4	39.3
T_h, hot leg hot temperature, K ($^\circ$F)	1005 (1349)	2 (4)	1002 (1345)	1002 (1345)
T_c, hot leg cold temperature, K ($^\circ$F)	772 (929)	2 (3)	776 (937)	778 (941)
m, hot leg mass flow, kg/s (lbm/s)	5.65 (12.69)	0.2 (0.4)	5.5 (12.2)	5.5 (12.1)
T_{ht}, hot tubes temperature, K ($^\circ$F)	842 (1056)	3 (5)	822 (1020)	824 (1024)
m_t, tube bundle mass flow, kg/s (lbm/s)	12.9 (28.3)	.02 (.04)	14.5 (31.9)	13.3 (29.3)
m_t/m, recirculation ratio	2.28	.08	2.62	2.43
f, mixing fraction	.81	.02	.96	.99

Table 9. Results for Leak 1

Result	2.8 kg/s leak rate	1.4 kg/s leak rate	0.14 kg/s leak rate	0.014 kg/s leak rate	scale-up case h5, 0 leak
tube heat loss, kW (BTU/s)	5304 (5028)	5193 (4923)	5037 (4774)	5030 (4768)	5047 (4787)
% tubes carrying hot gas	38.9	38.9	38.0	38.9	38.0
T_h, hot leg hot temperature, K ($^\circ$F)	1374 (2013)	1398 (2056)	1403 (2066)	1404 (2068)	1404 (2068)
T_c, hot leg cold temp., K ($^\circ$F)	909 (1177)	928 (1211)	942 (1235)	943 (1237)	942 (1236)
m, hot leg mass flow, kg/s (lbm/s)	4.7 (10.4)	4.4 (9.8)	4.2 (9.3)	4.2 (9.3)	4.2 (9.3)
T_{ht}, hot tubes temperature, K ($^\circ$F)	1040 (1413)	1046 (1423)	1046 (1423)	1044 (1420)	1046 (1423)
m_{ti}, hot mass flow entering tube bundle, kg/s (lbm/s)	11.5 (25.3)	10.4 (22.8)	9.1 (20.0)	9.0 (19.9)	9.1 (20.1)
m_t, hot mass flow in tube bundle above leak, kg/s (lbm/s)	9.1 (20.1)	9.2 (20.3)	9.1 (19.9)	9.0 (19.9)	9.1 (20.1)
m_t/m, recirculation ratio*	1.94	2.09	2.15	2.15	2.16
f, mixing fraction	.97	.93	.92	.93	.92
actual tube leak mass flow, kg/s (lbm/s)	2.94 (6.5)	1.42 (3.1)	0.14 (0.31)	.014 (0.031)	0
Mass flow entering leaking tube from inlet plenum**, kg/s	2.34 (5.15)	1.13 (2.49)	0.24 (0.53)	0.14 (0.30)	n/a
tube leak temperature, K ($^\circ$F)	942 (1236)	967 (1281)	999 (1339)	995 (1331)	n/a

* Based upon the hot mass flow rate above the leak location.
** If mass flow is less than the tube leak flow, remaining leakage comes from the outlet plenum. If mass flow is greater than leak flow, a portion of the flow passes the leak and continues to outlet plenum.

Table 10. Results for Leak 2

Result	1.4 kg/s leak rate	0.14 kg/s leak rate	0.014 kg/s leak rate	scale-up case h5, 0 leak
tube heat loss, kW (BTU/s)	4895 (4639)	5010 (4749)	5023 (4761)	5047 (4787)
% tubes carrying hot gas	39.8	39.8	38.9	38.0
T_h, hot leg hot temperature, K (°F)	1399 (2058)	1403 (2066)	1404 (2068)	1404 (2068)
T_c, hot leg cold temperature, K (°F)	931 (1216)	942 (1235)	943 (1237)	942 (1236)
m, hot leg mass flow, kg/s (lbm/s)	4.4 (9.7)	4.2 (9.3)	4.2 (9.3)	4.2 (9.3)
T_{ht}, hot tubes temperature, K (°F)	1046 (1424)	1041 (1418)	1042 (1420)	1046 (1423)
m_{ti}, hot mass flow entering tube bundle, kg/s (lbm/s)	10.0 (22.8)	9.1 (20.0)	9.1 (19.9)	9.1 (20.1)
m_t, hot mass flow in tube bundle above leak, kg/s (lbm/s)	8.9 (20.3)	9.0 (19.9)	9.1 (19.9)	9.1 (20.1)
m_t/m, recirculation ratio*	2.02	2.13	2.15	2.16
f, mixing fraction	.96	.95	.93	.92
actual tube leak mass flow, kg/s (lbm/s)	1.39 (3.0)	0.14 (0.30)	.014 (0.030)	0
mass flow entering leaking tube from inlet plenum**, kg/s	1.12 (2.46)	0.30 (0.65)	0.19 (0.43)	n/a
tube leak temperature, K (°F)	1142 (1596)	1139 (1591)	1137 (1587)	n/a

* Based upon the hot mass flow rate above the leak location.
** If mass flow is less than the tube leak flow, the remaining leakage comes from the outlet plenum. If mass flow is greater than leak flow, a portion of the flow passes the leak and continues to outlet plenum.

71

Table 11. Boundary Conditions From SCDAP/RELAP5 Predictions of CE Plant

condition	high temperature case	low temperature case
hot leg inlet mass flow, kg/s (Lbm/s)	6.2 kg/s (13.6 Lbm/s)	8.25 kg/s (16.5 Lbm/s)
hot leg inlet temperature, K ($^{\circ}$F)	1315 K (1907 $^{\circ}$F)	1010 K (1358 $^{\circ}$F)
T_{ct}, secondary side (sink) temperature (outside tube bundle) K ($^{\circ}$F)	875 K (1115 $^{\circ}$F)	750 K (890 $^{\circ}$F)

Table 12. Results for a CE Plant Steam Generator

Result	High temperature case	Low temperature case
tube heat loss, kW (BTU/s)	11008 (10426)	9286 (8796)
% tubes carrying hot gas	36.9	46.0
T_h, hot leg hot temperature, K ($^{\circ}$F)	1284 (1852)	985 (1313)
T_c, hot leg cold temp., K ($^{\circ}$F)	939 (1230)	777 (939)
m, hot leg mass flow, kg/s (lbm/s)	12.5 (27.5)	16.5 (36.4)
T_{ht}, hot tubes temperature, K ($^{\circ}$F)	1114 (1545)	871 (1108)
m_{tl}, hot mass flow entering tube bundle, kg/s (lbm/s)	17.9 (39.5)	27.0 (59.6)
m_t/m, recirculation ratio*	1.44	1.64
f, mixing fraction	0.58	0.64